Technology
Challenged

Technology Challenged

Understanding Our Creations
Choosing Our Future

Miguel F. Aznar

KnowledgeContext
Santa Cruz, California

Published by KnowledgeContext
800 Brommer Street, Suite 28, Santa Cruz, CA 95062
www.KnowledgeContext.org

Printed in the United States of America
on recycled paper

First KnowledgeContext edition, 2005

Aznar, M. F. (Miguel Flach), 1964 –
Technology Challenged: Understanding Our Creations
& Choosing Our Future / Miguel F. Aznar

Includes bibliographical references and index

1.Technology
2.Technology–Social Aspects
3.Technology–History
4.Technology–Risk Assessment

Library of Congress Control Number: 2004099558

ISBN-10: 0-9763858-0-5

Contents

Introduction

It might be a familiar progression, transpiring on many worlds—
a planet, newly formed, placidly revolves around its star;
life slowly forms; a kaleidoscopic procession of creatures evolves;
intelligence emerges which, at least up to a point,
confers enormous survival value; and then technology
is invented...In a flash, they create world-altering contrivances.
Some planetary civilizations see their way through,
place limits on what may and what must not be done,
and safely pass through the time of perils.
Others, not so lucky or so prudent, perish.
— Carl Sagan

The time of perils has already begun. This book offers a tool for navigating them. It is a personal tool, one we might use in our everyday choices. And it is a global tool, one that could help our civilization survive.

Just how did we get into this situation? Expanding on Carl Sagan's description: For most of the earth's existence it has hosted organic life. Single cell life evolved into multicellular plants and animals. Some animals started to use simple tools—from otters cracking open shellfish with rocks to chimpanzees dipping into termite mounds with sticks—but another animal went farther. It used tools to create even better tools. With spears it hunted. With sewn animal hides it survived the cold. With plows it created surplus. With tablet and scribes it recorded information. These tools led to the printing press, microscope, steam engine, telephone, airplane, and computer. We call these things *technology*, and for a million years they have been transforming our environment...with ever increasing power, costs, and benefits.

World War II saw the creation of one of our most potent contrivances so far: a weapon based on a nuclear chain reaction. Not as well known as the atomic bomb was a possible side effect of its detonation: a second chain reaction that might incinerate the earth's entire atmosphere. While the *nuclear* chain reaction was limited by the amount of radioactive fuel contained in the bomb, the *atmospheric* chain reaction would be limited only by the amount of oxygen cloaking the earth.

What to do? Scientists building the first atomic bomb estimated the risk to be as high as three chances in a million. They weighed the costs and benefits…to themselves, their country, and their planet. They considered the objective calculations and their subjective values before proceeding with the detonation.

The atmosphere did burn, but only in proximity to the explosion. There was no atmospheric chain reaction. Had there been, we would not be here to write about it. The atomic bomb is hardly alone in leveraging human power to a perilous height:

- One of the oldest tools, knives in the new form of box cutters, were used to hijack airplanes, which were used as suicide weapons in the U.S. on September 11, 2001. Technology gives individuals breathtaking power.

- The Institute for Biological Energy Alternatives synthesized a completely new virus in just two weeks. Unlike a bomb, a virus does not explode once, but can multiply and spread. Severe Acute Respiratory Syndrome (SARS) circled the globe in a matter of days.

- SARS was neither highly infective nor highly deadly, but Vector, the USSR's secret bioweapons laboratory, genetically modified diseases to be both. One product was a smallpox virus designed to be resistant to all known treatment. Scientists also worked toward viruses and bacteria that would degrade the human immune system or modify behavior.

- Electric probes implanted in the brains of rats have demonstrated rudimentary "mind control," foreshadowing the day when entertainment may become as immersive as video games and as addictive as brain-chemistry-altering drugs. What effect might "digital methamphetamine" have on society?

- Nanotechnology, the technique of creating objects on the molecular scale, would become more efficient if it could create microscopic robots—"nanobots"—that could, in turn, create more nanobots, which would create still more nanobots. Factories would then have workers that could also make more workers. But as Mickey Mouse discovered in *The Sorcerer's Apprentice* when his magic broom replicated itself without limit, a self-reproducing tool can quickly escape our control.

There are good reasons for almost any technology. Even though they can be used as weapons and can transport disease, airplanes provide tremendous benefits. Those developing a drug-resistant strain of smallpox must have believed that presented more benefit than cost, at least for them. The question is not, "Should we have technology?" Obviously, we have it and—barring catastrophe—we will have much more of it. The question is, "How do we evaluate it?" Unfortunately, the most compelling argument for a considered, critical approach would be a spectacular disaster…and that could exterminate us.

We live in an increasingly interdependent world and,
due to the progress of technology, our power over nature
has increased by leaps and bounds. Unless we use that
power wisely, we are in danger of damaging or destroying
both our environment and our civilization.
— George Soros

Evaluating technology is not just about saving the world, but is part of our everyday lives. Our education, work, health, and recreation choices pivot on technology. What kind of car is best for you? Which software should you buy? Should you take the new drug your doctor prescribed? Do you write your senator to support or oppose a missile defense shield or cloning? What jobs will technology move offshore...before it renders them obsolete? Billions of people making a thousand billion choices, aggregating—like raindrops building to a flood—to transform the earth. We did not get from stone tools to genetic engineering without choices.

Key to our individual and collective future is how we make those choices. But if we are to understand and evaluate technology, we face a monumental problem: technology is complex. There is more of it than we can fully understand and, beyond that, it is changing faster than we can keep up. Stone tools changed little over thousands of human generations, but modern technology changes radically within just a single human generation. Personal computers, cellular telephones, medicinal drugs, and weapons systems render themselves obsolete ever more quickly, and this trend continues to accelerate. So the small percentage of technology that any individual can fully understand is becoming smaller every day.

In our highly specialized world, even experts rely on experts. The auto mechanic fixes the car of the computer technician who fixes the computer of the mechanic. We rely on expert's reviews, take our doctor's advice, listen to our friends, and then make gut decisions. And, while our children exhibit an amazing facility for adapting to and using new technology, they are no better prepared to evaluate it. Schools teach them that technology is no more than computers, and that knowing how to operate them is equivalent to understanding them. Learning which buttons to push is no substitute for the ability to evaluate.

In an era of rapidly changing technology, studying the details of what has already been invented is like driving a car while craning out the window and staring down at the blur of asphalt. This is a dangerous way to drive. Learning to operate current tools is important for many occupations, but in order to plan ahead we need a grasp of the timeless patterns that have held true for stone tools, plows, computers, and genetic engineering...and may well continue to hold for future technologies. This is not about how to design a computer or repair a car, but about discovering a big picture that puts these technologies—all technologies—into context. It is technological literacy.

> There is a major difference between
> technological competence and technological literacy.
> Literacy is what everyone needs.
> Competence is what a few people need
> in order to do a job or make a living.
> And we need both.
> — William Wulf

This form of literacy changes how we perceive technology. Do we treat it as some foreign and strange thing that "experts" create and direct us to use? Or, quite the opposite, do we create a relationship with technology, putting it within our understanding and influence? Even a computer engineer or auto mechanic may sometimes take the first view, seeing the other as an expert whose technological domain is completely foreign. Technologically literate people take the second, more powerful view.

This "big picture" contextual view of technology is precisely what we need to navigate these perilous times—both on a personal and a societal level. To gain that view, we need to figure out what is true for many technologies, even those not yet invented. It is in our nature to seek out the patterns around us—all a part of finding our place in the Universe. We look for patterns in technology much the same way we would look for them in anything...by asking the right questions.

Overview

> Being technologically literate is knowing what questions to ask.
> — Ira Flatow

To understand and evaluate technology, we ask nine questions (each the topic of a chapter) and seek enduring answers. We start by building a foundation: In chapters one through four, we examine technology's *identity*. Once we identify technology, we analyze how it *changes* in chapters five, six, and seven. Together, these seven chapters give us a foundation for its *evaluation*, which we do in chapters eight and nine.

1. **What is technology?** Since we are looking for patterns that have long been—and will continue to be—true, we cannot define technology as just the latest inventions. We need a much broader definition. In Chapter One, we try several, including *tools that extend our abilities*. A definition that includes history—even all the way back to stone tools—may endure into our future.

2. **Why do we use technology?** In Chapter Two, we uncover a few answers that apply to most of what we have invented. Our desire to communicate, for instance, has been satisfied by writing, paper, printing press, pencil, radio, telephone, television, email, and "instant messaging."

3. **Where does technology come from?** In Chapter Three we look at environments conducive to the birth of technology. One pattern we find is that *denser populations* gave us more chances to encounter and build upon each other's inventions, speeding up progress. This might help us understand why technology changed slowly for thousands of years but rapidly now.

4. **How does technology work?** All humans use technology, much of it so simple—a hammer or shovel, for instance—that we do not even think about this question. Perhaps surprisingly, there are simple patterns that hold for a variety of technologies. One pattern we uncover in Chapter Four, *feedback* and *correction*, explains how computerized thermostats, robots, and other complex technologies work.

The four questions comprising the first four chapters focus on the "identity" of technology, but "change" is what makes technology a pressing issue, and we address it in the next three chapters:

5. **How does technology change?** While it is common knowledge that computers double in power every couple of years, few know that this *exponential growth* curve started back when computers were made of electromagnetic relays and vacuum tubes...or that mechanical clocks followed a similar curve of improvement beginning in the 1300s.

6. **How does technology change us?** It is not just the technology that changes. We change in response to it, just as any living thing adapts to its environment. In Chapter Six we find that technology has affected how we work, live, and perceive our world. Machines have displaced workers and created new jobs, pushing us up a pyramid of work, which often requires more thinking and less brute strength.

7. **How do we change technology?** Just as technology changes us, we change it. While Chapter Five looks at patterns of change intrinsic to technology, Chapter Seven looks at the ways that people influence it. As inventors, managers, investors, leaders, teachers, and in many other capacities, our choices and decisions guide technology. In a sense, humans form much of the environment in which technology either survives or becomes extinct.

8. **What are technology's costs and benefits?** In Chapter Eight, we search for patterns in the tradeoffs we make with technology. One tradeoff we examine: the more a technology enables us, the more we become dependent upon it. This was uncomfortably clear as we approached January 1, 2000 and worried about the Y2K bug, which had the potential to cause many millions of computers to malfunction. Computers are so useful that we have come to depend on them to schedule our factories, operate antilock brakes on our cars, and keep track of our bank balances.

9. **How do we evaluate technology?** In Chapter Nine, the second of our two chapters on *evaluation*, we draw on sociology and psychology. Countries, corporations, religions, clubs, families, and individuals bring their own values to bear when evaluating costs and benefits. For example, the values of Afghanistan's Taliban regime labeled Stinger missiles "good technology" and TV satellite dishes "bad technology."

These nine questions fit into the categories Identity, Change, and Evaluation (from which we get the acronym "ICE-9") as shown in this diagram:

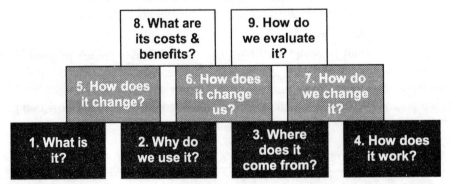

The ICE-9 questions are a honeycomb or structure of cubbyholes into which we can place new things we learn about technology. Asking those questions about technologies we encounter (directly or through TV, radio, newspapers, magazines, or books), we find patterns that hold true for many technologies. And these patterns form a context.

How might we apply these questions to a technology we return to later in this book, radios in North Korea? First, the background: any radios that can be tuned to frequencies other than the one carrying official broadcasts must by registered with the government. The tuners are soldered into place and police make surprise inspections, looking for tampering. Information is so tightly controlled that defectors are surprised to find that South Korea is more prosperous than North (which has had widespread starvation) and that U.S. donations of rice are not subservient gifts of tribute. Combating this dearth of information, a group in South Korea is smuggling in disposable radios. With ICE-9:

9. *How do we evaluate it?* The government of North Korea evaluates the radios in terms of their power. By promoting dissenting views, this technology is a threat to their control.

8. *What are its costs and benefits?* Like many technologies, radio offers tradeoffs between such goals as control and freedom. In this situation, radios subvert control and promote freedom.

7. *How do we change it?* Engineers design radios, activists distribute them, organizations fund them, and North Korean police hunt them.

6. *How does it change us?* Independent news sources heard over the radios change listeners' conception of reality: they discover that starvation is not normal and that their nation is not the world's most powerful.

5. *How does it change?* Electronic technologies, in particular, have become smaller and less expensive at an amazing rate, making disposable radios feasible.

4. *How does it work?* Many technologies can be characterized as either centralized or distributed. Unlike a large transmitter, the radios are highly distributed, so many could fail or be destroyed without affecting the rest.

3. *Where does it come from?* These radios come from specialization, designed by experts in microelectronics. Broadcasting, however, was an accident: radio was invented a century ago for one-to-one conversations where telephone wires could not be run.

2. *Why do we use it?* Communication is one of the oldest reasons we use technology and it still drives such devices as radios, satellites, cellular phones, and email.

1. *What is it?* Radio is a tool to extend our abilities, allowing us to hear something from far away. But the physical radio that we can touch is just the tip of the iceberg. Out of sight are systems of technical standards and networks of energy distribution and manufacturing just as important.

Technology takes on greater meaning when we understand its context. Oblivious of that context, many are satisfied to simply use technology, ignoring their relationships to it and its relationship to our environment. Dams also illustrate this point.

Looking down on the earth from space, some of the largest, most visible technologies are hydroelectric dams. Invisible from that high perch, however, is how interconnected the whole system is. Salmon feed in the oceans, enriching their bodies before returning to their spawning grounds. Unless dams block them. Salmon farms provide a spawning area below the dams, addressing the dwindling salmon population. But not a related problem: before creation of dams and farms, upstream bears, eagles, bobcats and many other animals ate salmon, and then fertilized inland trees with phosphorous and nitrogen from the ocean. Trees evolved over eons to thrive on that fertilizer, one of countless relationships now affected by technology.

But change is nothing new. Primitive tools changed how humans hunted, sheltered, and clashed. They changed the environment in which we evolve so, naturally, different traits became the most survivable—for instance, our ability to create and use tools. Medical technology, including antibiotics, has changed the environment in which viral diseases are competing to survive, helping to evolve antibiotic-resistant viruses. It has also extended human life, giving us time to philosophize or invent yet more technology.

> ...there could be a crucial hurdle
> at our own present evolutionary stage,
> the stage when intelligent life
> starts to develop technology.
> — Martin Rees

That we will change salmon, trees, viruses, and ourselves is inevitable. And, as technology advances, we will have greater power to cause change. The open question is whether we will effect those changes with a myopic view of the technology and its most immediate application, or with a view of the grander patterns.

We opened the Introduction with a quote from Carl Sagan cautioning us about the power and danger of our "world-altering contrivances." The danger comes from blindly embracing or rejecting technology—rather than influencing our world based on understanding and evaluation. Creating an intentional future is a collective process, and it is our hope that you, and those you pass this book along to, will find this approach useful.

Chapter 1

What is Technology?

Technology is a gift of God.
After the gift of life it is perhaps
the greatest of God's gifts.
It is the mother of civilizations,
of arts and of sciences.
— Freeman Dyson

The Hawaiian bobtail squid would be easy prey on bright moonlit nights if it cast a shadow. But it does not. Instead, the squid projects simulated moonlight on the ocean floor where predators wait. How does a squid extend its abilities to include shining like the moon? It gathers and eats bacteria called Vibrio fischeri. These communicate among themselves with chemical signaling molecules so they know how many of their peers have gathered, and when their population hits a critical density, or quorum, they glow. The squid packs these glowing bacteria into an organ with shutters, lenses, and colored filters so that it can simulate a wide range of moonlight, keeping the squid virtually invisible to predators. Does ingesting and using luminous bacteria qualify as using technology?

Vibrio fischeri have cousins named Vibrio cholerae, the water-borne bacteria that cause cholera. While the Hawaiian bobtail squid shines light with the help of Vibrio fischeri, the Vibrio cholerae bacteria actually change their environment. They enter the human stomach when infected water is consumed. At first it might appear that they are doomed for, unlike the benign bacteria found in healthy stomachs, Vibrio cholerae are killed by human digestive acids. Only one in a million survives. The survivors attach themselves firmly to folds in the lining of the small intestine and then inject a bit of toxin. The stomach's reaction to this threatened tissue damage is to flush the area with water, diluting the acid, washing away the other bacteria, and leaving the invader still clinging tightly. The Vibrio cholera procreate and, evolved to avoid putting all its eggs in one stomach, some ride the newly created river—diarrhea—in search of new hosts. All this flushing water dehydrates the human host and, untreated, cholera can result in death within hours. Is Vibrio cholerae acting as a technology because it changes its environment?

Is a sea otter smashing shellfish with rocks using technology? A chimpanzee smashing open nuts with rocks? A crow dipping for insects with sticks? Or a beaver damming streams to form ponds? Does instinctual use count? Or is being able to manipulate and share information about their tools—and being aware of these processes—necessary? It all depends on how we define technology.

The root meaning of technology, from Greek, is the study of a craft or art. John Lienhard, radio host and professor of both engineering and history, suggests that our species should not be called homo sapiens (the wise ones), but homo technologicus (those who use technology). He defined technology as "the knowledge of making things." In his book *The Technological Society*, Jacques Ellul defined technology in relation to art and science:

Art is concrete & subjective

Science is abstract & objective

Technology is concrete & objective

In this chapter, we explore several slightly more specific and practical definitions First we consider "any tool that extends our abilities," seeing how levers, pole vaults, and the Space Shuttle fit. Then we follow a story from one kind of rock that became important more than 2500 years ago to another kind of rock that has completely transformed our world in the past half century. Those two rocks and several technologies in between extended our ability to conduct commerce, which illustrates our second definition: "systems of tools." Homer's Iliad and the phenomenon of software piracy bring us to a third definition of "information as technology." There is no universally accepted and timeless definition, so we test our proposals against a variety of inventions and developments to see if they seem to make sense. In the last section we show why "applied science," although found in some dictionaries, comes up short for our purposes.

Definitions of technology help us to decide where to look for patterns. Too broad a scope may have few or no patterns that span it. Too narrow a scope hides patterns. Something true for televisions alone, for instance, is not nearly as valuable as a pattern common to prehistoric implements, agricultural devices, industrial machines, computer equipment, genetic tools, and even less tangible things, such as monetary systems. We want a tool for understanding and evaluating the technology of the future, so we look at technology of today and yesterday to get a feel for just what technology is.

Tools that extend our ability

How high can you jump? The Olympic record for the high jump is about eight feet. If you allow a simple technology like a pole, the record vaults to nearly 20 feet. Suppose you took a very large, hollow pole, and fill it with rocket fuel, add control systems, and provide a pressurized control module on top. Then, the record increases to nearly 300,000 miles with a trip around the moon.

Of course looping around the moon is not an Olympic event, but it does show that technology, perhaps by definition, extends our abilities. Testing out this definition, we will range from the first lever all the way to the bicycle and the Space Shuttle.

Unlike the moon rocket, the pole vault is simply a lever. Levers were in use long before Archimedes described them in 260 BCE, but he gets credit because his is the earliest known description. Long, long ago after a storm knocked down trees, one of our ancestors may have climbed atop one of them. With a lucky arrangement of trees, that stunned person would have lifted a massive tree off the ground.

The right arrangement involves three trees: *lever, fulcrum,* and *load.* The *load* tree lies atop one end of the *lever* tree, which lies across the *fulcrum* tree and extends up into the air. If the lever extends far enough from the fulcrum, the small force of the person's weight will lift the much greater weight of the load. Another reason Archimedes gets credit for the lever might be his memorable proclamation: "Give me a lever long enough and a fulcrum on which to place it, and I shall move the world." Long before television, he understood sound bites.

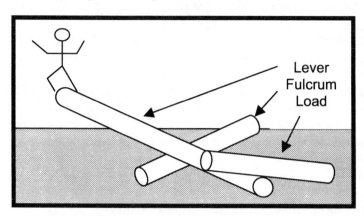

If the fulcrum is near your end of the lever, you find it hard to push the lever, but the other end moves much farther than does yours. Consider the garden rake left lying so that you step on the tines, propelling the long wooden handle towards your forehead at speed far greater than the descent of your foot.

If the fulcrum is near the far end of the lever, you find it easy to push the lever, but the other end moves much less than does yours. Consider a hammer turned around to pull nails out of a board. The claw end of the hammer moves an inch or two with great force while the wooden handle that you grasp moves six inches or more with less force. Depending on where you place the fulcrum, you trade force for distance or distance for force.

We refined trading force for distance with the bicycle. Equipped with multiple gears, it allows you to crawl up a steep slope or speed on a level surface. The bicycle incorporates levers into the crank arms, which connect the pedals to an axle. Chain and gears connect this to the rear wheel. The bicycle is a simple, yet highly efficient, technology that extends our ability to move. With a bicycle, a person can cover 200 miles in less than half a day, or ride across the U.S. in less than 10 days. Well, not a typical person, but some do take their recreation to these extremes.

A couple of bicycle mechanics named Orville and Wilbur Wright combined the ancient technologies of lever and wheel with a newer one, the airfoil, to make a practical airplane. With some significant upgrades (such as rocket propellant), the airplane became the Space Shuttle attached to the modified pole we described at the beginning of this section.

Tools that extend our abilities is a broad a definition, and would include television, which extends our ability of seeing far distances, and the DVD player, which extends our ability to see to times past. This definition can also include de-

structive ends. On September 11, 2001, a few violent people used a very old technology (knives…well, technically, box cutters), to take control of another technology (four large jet airplanes). The airplanes, with their load of refined aviation fuel, caused far more damage than a band of prehistoric terrorists wielding knives could have. Even a drunk driver, erratically maneuvering a ton of steel and glass, has his or her ability to do harm greatly extended.

If we think of a tool as an isolated object—such as an airplane, car, or television—then we are still missing something essential. Take almost any technology away from its infrastructure and it will fail. A car transported back in time 500 years would have few, if any, suitable roads, no source of gasoline, no source of replacement parts, and nobody would know how to operate or repair it. This has been a stumbling block in developing and deploying hydrogen fueled cars: no infrastructure of refueling stations. Similarly, a television in the 16th century would be useless. So technology must be more than individual objects; it must also include systems.

> In addition to tools and devices, we should include systems and methods of organization…Any collection of processes that together make up a new way to magnify our power or facilitate the performance of some task can be understood as a technology.
> – Al Gore

Systems: The Intangible Levers

A very special stone was discovered in the kingdom of Lydia (now Turkey) in 550 BC. A naturally occurring mineral containing silicon, this stone, called a touchstone, could reveal the purity of gold. As a result a new technology was invented. The "new technology" was not the stone itself, but the knowledge surrounding its use:

- Rubbing pure gold against a touchstone made a yellow mark.
- Rubbing gold diluted with silver made a white mark.
- Rubbing gold diluted with copper made a red mark.

This tool made possible a system of money by extending our ability to ascertain quality. The government minted gold coins imprinted with their guarantee of value…which could be tested by weighing the coin and rubbing it against a touchstone. Money—when trusted—extends our ability to trade. Think of it as an intangible lever.

Can you imagine trading without money? Suppose you wanted to trade goats for corn, but the person who had the corn you wanted was uninterested in goats? You would have to find someone who did want goats and would trade something of interest to the corn seller. Money could buy anything for sale, and, unlike goats, money did not become sick or die on your way to trade. While you could say that the coin, itself, is the technology that extends our ability to conduct commerce, the coin is just one component of a system—one that includes the government guarantee, touchstones, moneychangers, and knowledge among those who would accept it as payment.

That system became more sophisticated with the check or credit note, invented in 14th century Italy. It allowed international trade without packing along large amounts of money. Being at the center of Mediterranean trade routes, Italy harbored the banks that issued credit notes. Merchants purchased notes that the bank guaranteed could be exchanged for a set amount of a foreign currency in a specific city. On the dangerous roads and sea routes that the merchants traveled, robbers were interested in goods and money, not written notes.

The importance of the system, including knowledge and trust, is much more important with credit notes than with coins, which could presumably be melted down into something of value outside the system. The credit note and paper currency relied on information. Who issued it? How much is it worth? What are the terms for its redemption? The U.S. dollar continues to display signatures of government authorities and the assurance that it is good for all debts, public and private.

The system becomes even more sophisticated in credit card technology. Predicted in the 19th century novel *Looking Backward*, but made practical in the 1950s, the credit card allowed us to trade without carrying large amounts of cash or finding someone who would trust our check. Information from the credit card (the number) and the transaction (the amount) are transmitted to a central computer, which keeps track of credit limits, spending patterns, and stolen cards.

"Smart cards" further extend our ability to conduct commerce by carrying all that information on the card itself. Insert a smart card into an automated teller machine and "load" it with money from your bank account. What makes a smart card "smart" is an embedded silicon microchip that stores encrypted information about how much money you have transferred from the bank account to the card. You can then, for example, insert your card into a soda machine, and if there are enough funds on it, you will get your drink.

Money, in the form of encrypted information, is transferred to the soda machine, leaving less on your card and preventing you from spending the same money twice. Periodically, the information from the drink dispenser goes to a clearinghouse computer (which could be located anywhere on Earth) that credits the owner's bank account with the amount you spent. A credit card, by contrast, must *immediately* contact a central computer every time it is used, which can be slow and, for very small transactions, relatively expensive. By coincidence, the touchstone (the first tool extending our ability to conduct commerce) and the smart card (the most recent tool to do so) both contain silicon.

Money, checks, credit cards, and smart cards are all *systems* that extend our abilities. Beyond monetary systems are economic, legal, and business systems that also extend our abilities. Public and private organizations have countless linkages determining how they interact and cooperate. Look at a part of these systems in isolation and it will not make sense because its environment defines its behavior. And that environment is the system. We cannot understand technology without understanding its context.

Let's come full circle to the touchstone. The touchstone is not an invention—it is a naturally occurring mineral containing silicon. Yet the system of information surrounding its use (such as how to interpret the results) could be called technology. So, could information alone be considered technology? If so, the invisibility of information will make it harder for us to recognize new technologies that are largely or completely comprised of it.

This is not a new concept. As noted at the start of this chapter, the root meaning of technology, from Greek, is the study of a craft or art. In other words, it is the knowledge someone has of a practice, perhaps making pottery or sailing ships. Perhaps technology, then, is not just the tool that extends our abilities, but the whole system of tool and information about the tool.

One cannot really understand [technology]
without an understanding of the
roles, incentives, skill, and behaviors
that define its use.
– L.G. Tornatzky

Information: The invisible ingredient

Take away the information that surrounds the physical artifacts we call technology, and they don't work. That information specifies how to operate, manufacture, and repair, and ranges from ancient human techniques to modern computer code. You cannot always see those instructions, but knowing what to do is a critical component of technology.

Before Homer committed the story of the Iliad to writing in the 9th or 8th century BC, it was a song that included technological information, such as techniques for launching and landing ships. Recording information took a leap forward with writing, about 5000 years ago, and then again with the interchangeable type printing press almost 600 years ago.

In the 20th century, computers took on the role of manipulating and transmitting information. In fact, in a circular manner, they record information about the design of their successor computers. Engineers continue to design the next generation of technology by using the current one. At the end of the 20th century information sharing accelerated with the Internet.

Some technology is more information than material. For example, the cost to make one microprocessor is almost as much as the cost to make a thousand. The material cost of the silicon in a single microprocessor is nearly zero. The expense comes from manufacturing facilities, manufacturing setup, and research and development of the design. So the essence of a microprocessor is in how the few square centimeters of silicon is arranged into millions of tiny transistors. A lump of silicon is almost worthless, but a microprocessor sells for hundreds of dollars. And the information that separates the two is worth billions (one microprocessor manufacturer, Intel, spends that amount each year on research and development).

In some technology, the material surrounding the information is nearly irrelevant. Microsoft earns billions of dollars selling compact discs and lots of empty space in cardboard boxes. The value of their software technology, which can sell for hundreds of dollars, is in the information represented on the discs, not in the material of the discs, which is worth pennies. And even those pennies can be eliminated from the mix. Many companies allow purchase of their software by downloading it over the Internet. The buyer provides information (credit card authorization) and the company provides information (software instructions for the buyer's computer). No physical substance moves from the seller to the buyer. The technology is 100% information. And that invisible information we call "computer software" generates many billions of dollars in corporate revenues each year.

The fact that the technology can be 100% information also makes it easier to steal. Stealing 1000 cars is much harder than stealing one. However, making 1000 copies of pirated software is not much harder than making one. So, while the creators of information technology enjoy the economy of distributing their information, they also suffer from it.

This pattern of *technology as information* has a dark side, too. Weapons of mass destruction are sometimes classified as nuclear, chemical, or biological. Technologies in these categories have material and informational components. Those trying to limit proliferation of these have a much easier time controlling the material components because information moves so quickly and easily.

Years ago, publication of plans for an atomic bomb caused widespread concern. Fortunately, the critical materials are still hard to obtain. Nuclear technology was a huge advance in power, but if someone wished to use it—for instance, to blow up a city—that person needed refined radioactive materials. Although the collapse of the Soviet Union left some of its nuclear facilities vulnerable and countries such as North Korea are developing their own nuclear facilities, plutonium and similarly suitable materials are still much less accessible than information.

Chemical weapons use more commonly available materials, such as agricultural fertilizer. Publication or distribution of bomb recipes has made it easier for terrorists to create these. Although established terrorist networks can readily share this information, now any aspiring terrorist with an Internet connection can also easily obtain it, as we saw with the Oklahoma City bombing in 1995.

Biological weapons may be of greater concern than chemical because they can reproduce on their own. A bomb explodes once, but a plague can procreate and spread. Information about how to culture and reproduce disease agents (e.g. smallpox) is generally available. To avoid the danger, governments attempt to control the material component: strains of disease agents.

However, there is a more dangerous form of information concerning biological weapons technology: genetic engineering. In the near future, a disgruntled university student could take public information about how to modify microorganisms (e.g. viruses) and then use what will be common laboratory equipment to create a plague for which we have no protection or cure.

Information is a large and growing component of technology. It moves easily in books, on computer discs, and over the Internet. When it is part of a technology we consider "good," that speed benefits us tremendously. When it is part of a technology that threatens us, that speed undermines our control.

The trend appears to be toward information being more important than material in future technology. For example, nanotechnology (a new technology that we describe later) promises the capability of assembling almost any physical object from cheap, microscopic raw materials (e.g. the carbon atoms polluting our air). Companies could then sell the design for a toaster, bed, car, or almost anything. This information would be downloaded to a matter compiler, located anywhere, which would assemble the product, virtually out of "thin air." Today, that is still science fiction, but unless we become aware of this pattern of *technology as information*, we could still be hunting around for the tangible in a future that is all about information.

tech·nol′o·gy (tĕk·nŏl′ō·jĭ), *n.* [Gr. *technologia* systematic treatment. See TECHNO-; -LOGY.] **1.** Industrial science; systematic knowledge of the industrial arts. **2.** Terminology used in arts, sciences, or the like. **3.** Applied science. — **tech·nol′o·gist** (tĕk·nŏl′ō-jĭst), *n.*
tech′y (tĕch′ĭ), *adj.;* TECH′I·ER (-ĭ·ẽr); TECH′I·EST. [OF. *teche* a mark or quality (good or bad).] Unduly irritable; touchy. — **Syn.** See IRAS-

Excerpt from 1949 Webster's Dictionary. Note definition 3.

Not applied science

In the 1930s, a scientist at a dinner party used the back of a napkin to calculate whether a bumblebee's wings were large enough to lift it off the ground. The preliminary answer was that *if* the wings were rigid like those of an airplane, then the bumblebee could not fly. However, the bumblebee tilts and strokes its flexible wings quite unlike an airplane, so the scientist left the party to figure out how to take these complicating factors into account. In his absence was born the myth that, according to science, the bumblebee cannot fly.

The myth is popular to this day because it is an apparent flaw in one of the most powerful forces of the modern world. If someone had said that patterns in tealeaves deny the bumblebee's ability to fly, how many friends would you pass that on to? For centuries, science has been the world's leading source of truth, so it should not be surprising that some, including the 1949 edition of Webster's dictionary as shown in the box above, define technology as the application of science.

We have plenty of evidence of this application of science. When we ride a bicycle, drive a car, or fly in an airplane, we are relying on engineers who relied on science. Science predicts how things will work, often more quickly and economically than waiting until it is built. For example, the Wright Brothers used a wind tunnel to experiment with designs for their airplanes. And today equations can replace many physical experiments. But there are two reasons this is a poor definition: (1) scientific understanding often follows the creation of a technology and, (2) when science is applied to developing technology, the process changes from science to engineering.

One example of science trailing technology: thousands of human generations chipped at stones to create wonderfully sharp knives before the laws governing fractures of solids were uncovered. Another example can be found in radios. The "crystals," vacuum tubes, and transistors that made the first three generations of radios work were accidental discoveries, not applications of scientific knowledge:

- Early radios were called "crystal sets" because the radio wave detector was a crystalline nugget of germanium, galena, or silicon. Getting them to work required probing the crystal with a wire until a signal came through and then keeping the wire pressed against that magic spot. This allowed electricity to flow in only one direction (rectification), but "crystal set" radios were used for years before the rectifying properties were identified, and they were not understood in a scientific sense until after the transistor was developed.

- The vacuum tube came from the incandescent light bulb, in which Thomas Edison had noted what he called the "Edison Effect," but saw little use for it. Others developed it into a rectifier and amplifier, indispensable components of radio, television, and computers.

- The transistor came from crystal sets. Why these minerals rectified electricity was not understood scientifically, but Bell Laboratories thought they could improve on the vacuum tube (which, like their light bulb forebears, consumed lots of energy and easily burned out). That refinement of germanium and silicon crystals into transistors with precisely controlled amounts of impurities inspired scientific research into semiconductors, which led to integrated circuits and the boom in electronics and computers.

Science did lead the way in the *discovery* of germanium, if not its use in electronics. When Dmitri Mendeleev presented his periodic table of the elements in 1871, there was a gap between silicon and tin. His scientific approach told him that even if nobody had yet discovered it, there must be an element to fill that gap. Calling the as-yet-undiscovered element "eka-silicon," Mendeleev accurately predicted its weight and properties well before 1886, when it was discovered in Germany and officially named "germanium."

A more recent example: in 2001 Bell Labs created transistors so small that each used just a single molecule, so 10 million would fit on the head of a pin. The director of quantum-science research at Hewlett-Packard, Stan Williams, remarked, "They don't have a clue how or why this works and I don't have a clue how or why it works either." IBM's director of physical sciences research, Thomas Theis, agreed: "It appears to be a very interesting result, but nobody, including the authors of the paper, seems to fully understand what is going on here." Sometimes inventing is easier than explaining.

A second problem with defining technology as applied science lies in science being abstract and technology being concrete. Applied science bridges that gap, but it is only the bridge. The engineering process incorporates formulas and laws from science, but goes well beyond them in balancing costs and benefits. How strong does something need to be? How long do we have to test it? What are the costs of designing to far exceed the expected range of use? These are practical questions that have little do with science and everything to do with actually making something useful.

We are not done with the Hawaiian bobtail squid. The light from its luminescent bacteria is reflected by platelets composed of an extraordinary protein named *reflectin*. Scientists are studying that protein to figure out how it works, which may help engineers create microscopic optical devices. So, even if the Hawaiian bobtail squid is not using technology, it may inspire some. We can appreciate the importance of science in arming our engineers in their quest to create useful things, but we are better off without an applied science definition of technology.

I don't know who discovered water,
but I'm sure it was not a fish.
— Marshall McLuhan

Can you imagine trying to explain "water" to a fish? You couldn't point at water because, where fish live, it is everywhere. You have to stand apart from something to point at it. In the 21st century, technology is to humans as water is to fish. Opening the chapter with squid, two kinds of bacteria, sea otter, chimpanzee, crow, and beaver was a trick to get us to stand apart from technology and point at it.

What makes understanding and evaluating technology urgent is its rapid change, pointing to a future in which it will be even more powerful. Whatever your personal conclusions as to whether these or other animals use technology, it is clear that, so far, only humans have consciously changed it. Instinctual use allows tools to change only as quickly as instincts. Even imitative use, as chimpanzees and birds demonstrate, keeps tools relatively static. It is the dynamic nature of technology that makes it interesting. Carl Sagan would not have warned us of the "times of peril" had technology been frozen at the stage of stone tools.

But even as we collectively change technology, individually many of us are tricked into the myopia of equating computers and electronic equipment with technology. Those who can't see beyond those current specimens are swept along. While fish have the choice of fighting the current or going with the flow, humans have the further option of guiding its course…if we are aware. And awareness is what this chapter is about. It sets the scope for investigations to come about our relationship with technology.

The definition *tools that extend our abilities* is an important step beyond computers and electronics. Both stone tools and technologies not yet invented fit this definition. To prepare our eyes for that yet to come, we recognize the modern trend of technologies fitting within ever more complex systems.

Perhaps many of the inventions that can stand apart from modern systems were long ago invented. For modern inventions, survival of the fittest is determined within an environment of systems. Recognizing technology as *systems, the intangible levers*, we are more likely to spot future developments. Other trends suggest that *information, the invisible ingredient*, is becoming ever more important in and as technology. Designing, manipulating information, pays better than manufacturing, manipulating material. Nanotechnology could one day automate manufacturing, making it so inexpensive that what we care about is the information in the design of technology, not the material.

While our criticism of "applied science" might be seen as an exception, our purpose in this chapter is not to arrive at a single, universal, eternal definition of technology. Rather, it is to provide some thought-provoking answers to the question, to help you come to your own definitions. Each of the nine chapters in this book has a similar goal. The question that heads each chapter is nearly timeless, but the answers cannot be—technology changes too quickly. Picking a single best answer would be no more than an intellectual exercise, so, instead, we offer context as a platform from which to launch.

Chapter 2

Why do We Use Technology?

…a new communications technology…
allowed people to communicate almost instantly
across great distances, in effect shrinking
the world faster and further than ever before.
A worldwide communications network…
it revolutionized business practice, gave rise
to new forms of crime, and inundated its users
with a deluge of information.
– Tom Standage

The telegraph was unlike anything that had come before. Suddenly news could travel as dots and dashes of Morse code through a cable in the Atlantic Ocean between Europe and America. Letters bobbing for weeks on steamships could be replaced by speed-of-light conversations. Harnessed lightning replaced paper, changing business and crime. With improved communications, some predicted the end of misunderstandings between countries and the end of war. As the 19th century came to a close, the unique technology of the telegraph spread its cables like a giant octopus covering the world.

But the telegraph was not unlike anything that had come before. Other technologies had earlier improved and even transformed communication. And others, such as the Internet, would follow. As mind boggling as it was to move from handwritten letters to invisible pulses of electricity, using technology to communicate was familiar. Writing had transformed communication, as had papyrus, cotton paper, wood paper, printing, and printing with interchangeable type. The telephone replaced dots and dashes with voice (and seemed so fantastic that telegraph companies rejected the idea). Radio replaced wires, and satellites extended radio's range to circle the globe. The Internet added data in the form of text, graphics, and video to the voices we could already send. Cellular telephones made sure we could connect nearly anytime and anywhere.

What is yet to come? Technologies as baffling to us as the telegraph was to those living in the 19th century, and used for many of the same reasons we have always used technology. Finding something relatively constant in the torrential flow of technology is valuable in this period of rapid technological change. We picked a dozen categories for why we use technology:

1. Food
2. Shelter
3. **Communication**
4. Transportation
5. Commerce
6. Art
7. Religion
8. **Health**
9. **Entertainment**
10. **Organization**
11. Conflict
12. Exploration

In this chapter we give examples for the four categories in bold. You may think of reasons to use technology that do not fall into one of these categories. Or, you may find them too specific, and be tempted to generalize them into five or six, similar to the taxonomy of life.

But more important than the specific categories is the benefit of having *some* categories. Basic human needs and desires change little, and we can expect that future technology will simply find ever more creative ways to satisfy them. These categories—or whatever set you adopt—can be a template to place on unfamiliar technologies.

While this may temporarily blind us to a truly new purpose, handled carefully, it will help us past the marketing hype of new technologies. In most cases, that "completely new, does everything, unlike anything that has ever existed" innovation will satisfy one or several of these common reasons for using technology. Determining which needs it satisfies will help us find other familiar patterns (e.g. the printing press made it easy to print trashy novels; the web made it easy to publish trashy websites).

In the following sections, we illustrate each category with one or several technologies. Finding examples was easy—every newspaper or magazine article that mentions a technology includes some implied or explicit reasons for its use. Deciding which to include was not. A comprehensive list of all technologies used for a given purpose would be endless—we would have to include every technology in existence. And a ranked list showing only the most important technology in each category could be predictable and even dull.

So instead, we looked for the most entertaining illustrations for each category. Do not be disappointed if something as important as the printing press has been pushed out of the limelight by the "high-tech" cigarette, or if we spend more time on Entertainment than Communication. Neither is a claim of relative importance, but simply an acknowledgement that, elsewhere, the likes of the printing press have received "plenty of ink."

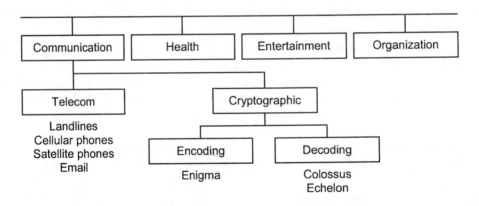

For thousands of years, kings, queens, and generals
have relied on efficient communication in order to
govern their countries and command their armies...
It was the threat of enemy interception that
motivated the development of codes and ciphers.
– Simon Singh

Communication

On a winter night in 1985, an Iraqi shepherd felt warmth coming from the hill he was sitting on. The surrounding slopes where his sheep rested were cold, so he was very curious. Digging into the earth on that remote spot 300 miles west of Baghdad, he found a warm metal tip connected to a machine. It was connected to Iraq's main telephone trunk line with Jordan. A nuclear cell powered it to transmit everything it heard to listeners unknown. Demolitions experts tried to open it, but it exploded, killing two.

In *Saddam's Bombmaker*, Khidhir Hamza reported that, "According to interviews the security people conducted with other shepherds and Bedouins in the area, helicopters with Iraqi markings had unloaded soldiers on the hill a few months earlier. They'd seen the soldiers digging on the hill, and even heard them talking in Iraqi slang." But those soldiers were not Iraqis. Few neighboring countries trusted Saddam Hussein, but Iraq was sure that Israel, alone, had the capability for this elaborate telephone-tapping operation.

Eavesdropping probably predates writing, but we have historical evidence for the use of secret writing shortly after the development of writing itself. In ancient Egypt, priests used Hieratic ("sacred writing") to keep communications secret. Cryptography, the science of encoding and decoding information, has made use of many technologies, and it has spurred the development of some.

Hidden beneath the rough, dark waters of the Atlantic German U-boats searched for Allied ships to sink. World War II German naval commanders were so confident of the imperviousness of their Enigma encryption machine that they regularly radioed orders to their subs at sea. But, within half a day, Britain could figure out where the subs were heading. How? British code breakers used Colossus, the first electronic computer (though some call the machine, built from 1500 vacuum tubes, a calculator rather than a general purpose computer).

This is how it worked. Enigma machines used a typewriter keyboard and electrical connections routed through several 26-sided wheels or rotors, which scrambled the letters that were typed. On the receiving end, another Enigma machine with identically wired rotors unscrambled them. When Germany suspected that their codes in their three-rotor machines had been compromised they then added a fourth rotor.

Even with information about the Enigma machines captured by Polish and French resistance fighters, England could not take a brute force approach to figuring out how the Germans wired up the rotors each month. Even if they could test 200,000 states each second, it would have taken more than 15 billion years, roughly the age of the Universe!

To overcome this challenge, England needed three things: its Colossus computers, human ingenuity, and human fallibility. The ingenuity was analyzing the encrypted messages for patterns in the German language, which could show through the encryption—even if so faintly that only a computer could detect it. The fallibility was the German practice of announcing each victory to every far-flung military unit in precisely the same language. This gave England multiple copies of a message, each encrypted differently by the same wiring of the rotors.

In the 21st century, our reliance on telecommunications is even greater and encryption has become a political issue. The National Security Agency in the U.S. works with Britain, Canada, Australia, and New Zealand to monitor and analyze global communication. Telephone, fax, and computer messages are intercepted by "Echelon" computers, which look for the signature of a terrorist plan or other security threat. Humans review the most interesting material once it has been filtered down from an immense number of intercepts.

Some European nations complain that this monitoring picks up business information, which is then shared with U.S. companies, giving them an unfair competitive advantage. As with any arms race, encryption has been improved to foil the Echelon monitoring, but U.S. law prevents export of any encryption system beyond a certain ability (presumably that level above which government computers could not decrypt).

Some communications are easier to monitor than others. The increasing use of cellular and satellite telephones is broadcasting more conversations into the atmosphere, but, as the Iraqis discovered, even "land lines" are not secure.

While the Internet and Web are capable of much more sophisticated applications, two of the most popular have been email and instant messaging—simple, quick, inexpensive communication. Future technology, however strange it may appear, may also satisfy this enduring human need to communicate.

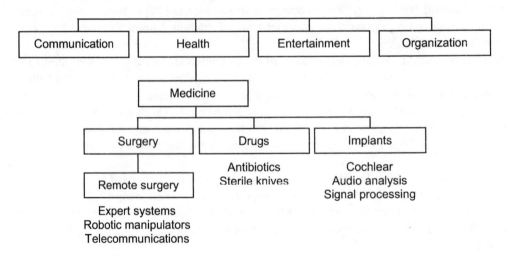

I felt as comfortable operating on my patient
as if I had been in the room.
— Jacques Marescaux, MD

Health

How could a surgeon operate in a room distant from the patient? Cameras transmit views of the patient to the surgeon and remote controls allow the surgeon to operate robotic manipulators. This technology was developed for surgeons in the same room as their patients because it can be slipped through tiny incisions, which are much less traumatic for the patient than holes big enough for the surgeon's hands. Another advantage is that relatively large finger motions can be translated to miniscule knife or probe motions, giving the surgeon much steadier and precise hands.

Why would a surgeon operate in a room distant from the patient? The surgery you need may have been studied by a local surgeon, but actually performed hundreds of times (successfully!) by a surgeon in another part of the world. The odds are better with the veteran…if the technology gives the surgeon a good enough feel.

The key to remote surgery is dividing the process into stages some of which involve only information, such as steps two and four below:

1. The patient is viewed by digital cameras
2. **Information** from them is transmitted to a computer screen
3. The surgeon views screen and manipulates computer controls
4. Those controls transmit **information** to robotic "hands"
5. The robotic hands interact with the patient

Our global communication network is good at transmitting information anywhere. As long as the cameras and robotic manipulators are in the same room as the patient, the viewing screens and controls are in the same room as the surgeon (and the system does not crash), then the distance does not much matter.

Computers can further change the motion of manipulators by incorporating the typical movements of recognized experts in each surgical area. These *expert systems* imitate the best practices, allowing them to be used even when the experts are not present. Eventually, this may go beyond minor modification of surgical movements with computers performing surgery on their own. An attending surgeon would switch on an "auto pilot," much as airline pilots commonly do today.

Still very expensive, computer-assisted surgery is not yet bringing the best of surgery to poor areas that lack any form of it at all. Do benefits once reserved for the few ever trickle down to the many? Well, in 1836, one of the richest people on earth died from something that, today, any pharmacy with antibiotics could cure.

At the time, germs—the invisible creatures that we so carefully sterilize from open wounds and surgical instruments with heat, alcohol and high-tech substances today—were not yet discovered. So Nathan Rothschild, an otherwise healthy 59-year-old banker, died of a simple infection from an abscess or boil—or from the surgeon's attempt to open it with a non-sterile knife.

Medical technology was primitive by current standards, and did not include the antibiotics that could have saved him. And current antibiotics are primitive compared to the eventual products of biotechnology.

Rothschild could not have imagined the reach of current medical technology. Cochlear implants are electronic devices placed in the inner ear, or cochlea, that bring hearing to the deaf. They convert sound waves into electrical impulses to stimulate nerve endings. For those with nerve damage in the cochlea itself, newer implants connect directly to the brain stem.

To test this technology, a cat in California has a brainstem implant for hearing and a person has a brainstem implant to control a cursor on a computer screen. That person suffered from a brainstem stroke and lost use of his hands, but he can still interact with a computer, which picks up his thoughts on wires that pass through his skull to the implant.

Advances borrowed from other areas—for example the sophisticated audio analysis and signal processing done by spies at the KGB, CIA, and NSA—could make future cochlear implants superhuman. Expect that the future will bring more and more technology to satisfy our quest for health because the consumers with the most disposable income have many of their other needs, such as food and shelter, already met.

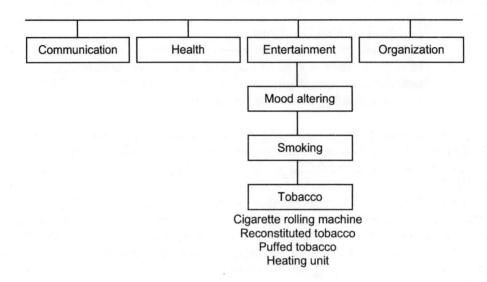

The constants all through the centuries
will be the same: wine, women, and song.
Other than that, life will be very different technologically.
– Phyllis Diller

Entertainment

Smoking tobacco, which we categorize as a form of entertainment (albeit, an addictive and dangerous form) has its share of technology. The cigarette rolling machine, invented in 1881, helped cigarettes eclipse pipes and chewing as the most popular form of tobacco. But as important as that mechanical technology was, chemical technology is the key to cigarettes' insidious power.

Here is how it works. Two types of tobacco go into most cigarettes: *reconstituted* and *puffed*. Puffed tobacco is made from tobacco leaves saturated with freon and ammonia before they are freeze-dried, which doubles their volume. Reconstituted tobacco is made from tobacco stems and parts of the leaf that cannot be used in puffed tobacco. These are pulped and then sprayed with hundreds of chemicals including nicotine, which is also found naturally in the leaf.

Nicotine is the most important chemical in cigarettes. Highly addictive, it diminishes appetite, affects mood, and can, at least temporarily, improve performance. Tobacco company laboratories developed chemical additives to improve delivery of nicotine. Ammonia, for instance, makes more of the nicotine vaporize when heated by the burning of the cigarette. Vapor can readily travel to the lungs, where the nicotine accompanies oxygen into the blood, which flows to the heart (which can speed 10 to 20 beats per minute with the first nicotine "hit" of the day), which pumps it to the brain.

In the brain, nicotine affects neurons, the nerve cells behind our thoughts and feelings. Specifically, nicotine mimics chemicals that neurons use to communicate with each other, over-stimulating the neurons with many false signals. By interfering with normal neuron communication, nicotine can alter mood, often pleasurably.

Since the brain, like most life, is highly adaptable, it accommodates to the over stimulation of neurons. As a result, when a smoker stops smoking, the brain initially perceives the normal level of stimulation that resumes as inadequate. The unpleasant symptoms the former smoker experiences are called "withdrawal," and they can be alleviated by resumption of smoking.

A Spanish historian noted the addictive nature of cigars in 1527. When science caught up with conventional wisdom and declared smoking dangerous to one's health, the highly profitable industry started working on a "safer" cigarette.

The first danger their scientists took on was tar, one of the many chemicals that smoking introduces into the lungs. They created filters, air holes that dilute the smoke with fresher air, and low-tar blends of tobacco to reduce the amount of tar going into smokers' lungs.

But they also reduced the nicotine, which smokers' brains were finely attuned to. Just as your brain can adjust your throw when the ball falls short of the basket, smokers' brains adjusted the puffing when the nicotine fell short of the "norm." By inhaling deeper, covering the air holes with fingers or lips, or smoking more cigarettes, smokers were able to get their accustomed nicotine levels. This also restored the previous levels of tar.

The Accord™ cigarette, introduced in 1998, is a "high tech" approach to safer cigarettes. The smoker inserts one end of a special cigarette into the microchip-based heating unit. Because the tobacco is not burned away into ash, the unit has a liquid crystal display (like those found on watches and calculators) to indicate how many puffs remain in the cigarette. After each pack of cigarettes, the smoker must recharge the batteries in the Accord.

The Irony of "Safe" Cigarettes

Heating tobacco, as the Accord does, rather than burning it, as conventional cigarettes do, produces no carbon monoxide or secondhand smoke. Carbon monoxide is also found at the tailpipes of cars, and can be fatal when a car is run in a closed area, such as a garage. Secondhand smoke has led to many state laws prohibiting smoking in public buildings and even at outside areas like theater lines or building entrances, where smoke might be drawn inside. And yet, manufacturer makes no health claims about the Accord. There is a good reason for this.

The technological problems with "safe" cigarette are dwarfed by the political problems, a pattern we will see with other technology. Tobacco companies worry that developing and selling "safer" cigarettes would be viewed by courts as an admission that other cigarettes are not safe. The lawsuits have stakes in the billions of dollars.

Further, tobacco companies are concerned about regulation by the U.S. Food and Drug Administration (FDA). Current tobacco products are exempt from FDA scrutiny due to a "grandfather clause" under which a new law does not affect someone (or something) that preexisted the law. But a new class of "safe" cigarettes might not fall under that clause. The irony is that the safest course for tobacco companies—if not for their customers—seems to be to avoid "safe" cigarettes.

In the movie *Sleeper*, Woody Allen plays someone cryogenically frozen and then thawed in the future. There, a favorite form of entertainment is touching a metal ball that makes the toucher feel good. Humans already spend lots of resources on feeling good, so a future technology that effectively and efficiently does that will be in demand.

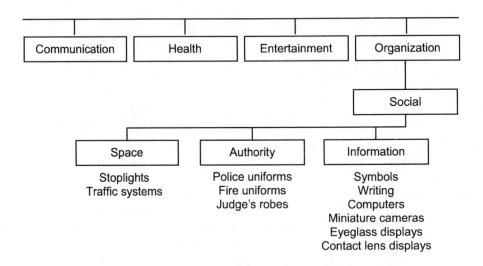

...social groupings larger than 150-200
become increasingly hierarchical in structure...
There must be chiefs to direct, and a police force
to ensure that social rules are adhered to.
— Robin Dunbar

Organization

Why would we need technology in order to organize? An answer comes, circuitously, through a story about the brain's neocortex. And we start with chimpanzees.

Take three chimps in the same band. Each chip is aware of his relationship with the other two and of the relationship *between* the other two. Who is dominant, who has done favors for whom, who can be trusted to repay favors, and who cannot? Chimp decision-making has been observed in the wild. Two chimps may team up to attack another chimp to steal food...unless the victim is near others who may come to his aid. Any chimp for whom the victim has performed a recent favor, such as grooming, is suspected of being a supporter.

Social animals keep track of their relationships with other group members and they also keep track of the relationships between them. In highly social groups, it is a matter of survival to know how others will interact.

As groups become larger, there are more relationships to track. With just two individuals there is just one relationship. With three individuals there are three relationships (shown as arrows below) and with four there are six:

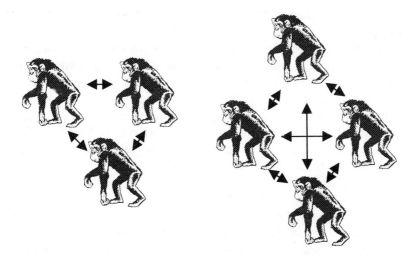

Metcalfe's Law

Chimpanzees in a tribe form a network. In a network, the number of possible one-to-one relationships is proportional to the square of the number of individuals. Doubling the tribe quadruples the number of possible relationships. Multiply the tribe size by three and the possible relationships multiply by nine.

People, telephones, computers, and railroads form networks, too. Robert Metcalfe suggested that the value of a network is proportional to the number of possible relationships, which is the square of the number of nodes. Those who accept this "law" as true are equating value with possible connections.

Network technologies such as telephones, fax machines, pagers, cellular phones, and email accounts have grown slowly at first, but accelerated suddenly once they reached some critical mass. The first person to own a telephone could do little with it, but today, having a phone is indispensable because it can make so many connections. When just a few university scientists used email, it had little value to most people. Today, many rely on email for both work and play.

A primate study has shown that the size of the brain's neocortex correlates with the size of the social groups in which that individual lives. Based on the pattern found in non-human primates, the human neocortex suggests a maximum group size of 147.8, or about 150. In a group of 150 individuals, there can be 11,175 such relationships, too many to memorize as a list, but not unreasonable to learn in context. A soap opera aficionado has no trouble remembering which of dozens of characters hate each other, for instance.

While that number of 150 still guides the size of clans, military "companies," fraternities, and church congregations, technology has steamrolled over it with cities. Technology concentrated people with agriculture and then united disperse populations with various communication technology. How do we cope with more relationships than we can remember?

We define relationships with technology. Stoplights and traffic systems define relationships with other drivers. Uniforms define relationships with police, fire, and medical personnel. A judge's black robes are symbolic of the system defining our legal relationships.

Symbolic language, writing, and computers help us manipulate, store, and transmit symbols of our relationships.

Still, there is stress from living and working with so many people. How can technology help our relatively unchanging biology adapt to our increasingly complex surroundings? Imagine knowing everyone you see. Unless you live in a small village, you would probably have to be augmented by technology. Technology will eventually be able to identify everyone you come in contact with, providing you information about whom they are related to, what their interests are, what they do for a living, and what their friends or former friends say about them. This information could be displayed on your glasses or contact lenses so that walking down a busy sidewalk, there would be no strangers.

Do we want village life with 10 billion people? Whether we want it or not, technology is enabling this level of familiarity. Already, glasses exist that project information on your field of view, like the heads-up display used by fighter pilots, helicopter gunship pilots, and the drivers of some cars. Attached cameras, wireless links to remote computers, and image recognition software exist and their commercialization is not far off. Once the glasses are reduced to contact lenses our remarkable familiarity with each other will appear quite natural.

To complete this utopia—or nightmare—we need only add "Big Brother" databases to store and cross-reference all this information. In 2002, the U.S. Information Awareness Office was formed to do just that in order to fight the war against terrorism.

As with each category we have touched on, it has been only a touch. Many other technologies satisfy our desire and need to organize. The web, for instance, organizes people through chat rooms and virtual communities. Terrorists are known to have used the web to organize their efforts. Humans are social animals so it seems inevitable that we would use technology to organize.

Why do we use technology today? For many of the same reasons we have always used it. Perhaps for the same reasons we always will use it. In our quest to understand and evaluate technology, "why do we use it?" is a powerful question because it so quickly categorizes even those things we do not yet understand.

Just as science has categorized life into taxonomy, we may categorize the reasons we use technology. The diagram below shows a standard classification (or taxonomy) of living things into five kingdoms, and then traces humans through the increasingly specific levels of phylum, class, order, family, genus, and species.

The diagram also shows how we might start to classify the reasons we use technology. We are not saying that technology fits into the classification of living things, just that a similar approach may help us understand why we use it.

The classification of living things has changed over time. It started with just two kingdoms: animal and plant. Fungi earned a third category and then science discovered life worthy of additional categories. Some scientists suggest a 6th kingdom (archaea, an ancient form of life that evolved separately from bacteria and blue-green algae), showing that classification does not lead to a single, obvious, and universally accepted answer.

As useful as our map may be, there is danger we might confuse it for reality, attempting to force any new technology into one or several of our categories. If it did not fit, we might ignore it, believing that anything outside a category is unimportant. Or we might miss an additional use for a technology simply because it satisfies some other need so well (e.g. a hammer as art).

Still, just because a tool can be misused, does not mean that it cannot be properly used. Our tool is, in a sense, itself a technology, one used for exploration. And, like all technologies, our categories are a double-edged sword.

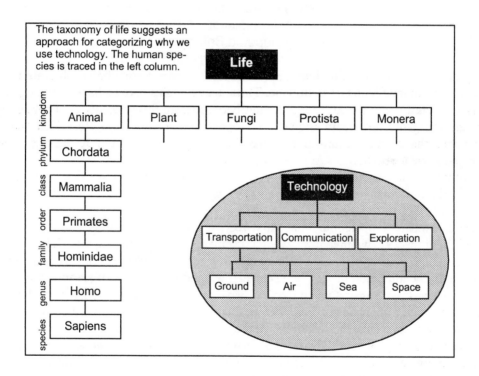

Chapter 3

Where Does Technology Come From?

An ape may on occasion use a stick
to beat bananas from a tree;
a man can fashion the stick into a cutting tool
and remove a whole bunch of bananas.
Somewhere in the transition between the two,
the hominid, or the first manlike species, emerges.
By virtue of his nature as a toolmaker,
man is therefore a technologist from the beginning,
and the history of technology encompasses
the whole evolution of man.
– Encyclopedia Britannica

Imagine a time when technology did not affect our lives, a time before the Internet, microprocessor, atomic bomb, and airplane. Those technologies take us back to the beginning of the 20th century, but we have to go farther back…before the automobile, telephone, light bulb, electric generator, steam engine, telescope, printing press, and clock. That takes us back before the 2nd millennium, but we have to go farther back…before the plow, wheel, lever, and stone tools.

That time, when technology did not live with us—*changing* how we live—predates history. It is hard to imagine 2.6 million years, so we compress it to just one year, ending today, illustrated on the next page.

Just as January 1 opens our imaginary year, we invent stone tools. Later that month, around about the 29th, we invent the wedge, useful for prying things apart. These technologies are enough to keep us busy—refining and improving, maybe even losing and rediscovering—until the afternoon of October 22, when we perfect the trick of starting and controlling fire. Or at least that was when we first leave enough evidence to convince later archeologists of our accomplishment (long before this we made use of naturally occurring fires). Good job, everyone. Take the rest of the year off. December will be busy.

Christmas Eve we invent the bow and arrow to hunt. Less than a week later, on December 30, we create wind musical instruments, but we save everything else for the next day. Before dawn on that last day of our imaginary year, we invent the plow and the wheel, so we're producing food surpluses and soon carting them about. Those food surpluses allow for specialization of labor, so before lunch we plumb our first bathroom and, at lunch, invent glass.

A minute before quitting time on December 31—4:59 PM—we find our bearings with the magnetic compass and decide there's more to do. By 8:30 PM, a mechanical clock tells us how late it's getting. By 10:41 it is quite dark, and we view the stars with a telescope. The steam engine and electric battery appear around 11:15. With less than half an hour left to go in the year, we shrink the world with telephone, automobile, and airplane. The power of the atom succumbs to our investigations at 11:48.

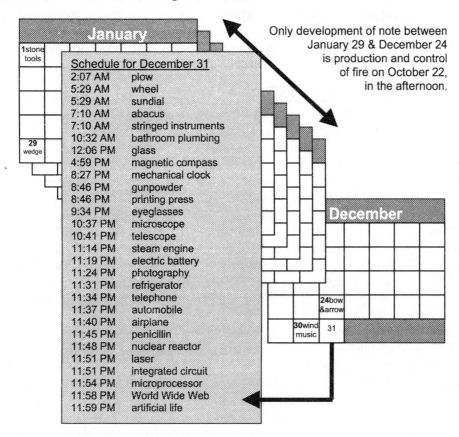

The epidemic spread of the integrated circuit and microprocessor start a few minutes later, leaving us just a couple of minutes to experience the World Wide Web. Then, in the very last minute we develop the ability to create artificial life, a 300-gene virus, assembled gene-by-gene in a laboratory. For ethical and moral reasons, we postpone its actual creation, but we do finish decoding our own genome.

Oh yes, in the final nanoseconds (each billionth of a second corresponding to about one of our real days), you start reading this book, which answers questions including, "Where does technology come from?" And to this question, we offer five answers. Technology comes from:

1. Other technology
2. Dense populations
3. Specialization
4. Plan or Accident
5. Protection.

Technology comes from *other technology*. In a sense, stone tools made possible all technology that would follow. While we cannot make a microprocessor with stone tools, we could not make one now if stone tools had not started the process. Stone tools had to be somewhere before the axes and spears, sewing needles, rope, pottery, fishing nets, and baskets. With stone tools, we were able to make better stone tools and also sharpen or carve wooden tools. With those, we made still better tools, slowly at first, but accelerating to and through the present day.

Even the idea of creating technology must have been easier to conceive once we started using stone tools. Of course the same is true for many technologies that extend our physical abilities to create and spark our thoughts of something similar but better.

Dense populations accelerated the pace of innovation near the end of our calendar. Early on, with tool-using humans grouped in small, dispersed bands, one invention had little chance of being seen by another inventor who might improve upon it. If one person invents the hand ax and another person is skilled at lashing things together, their combination could result in a conventional ax with handle (giving it the lever advantage we enjoy today). But, if those two people are isolated, neither benefits from the discoveries of others. No synergy. Progress is made only by repeated advances within each community.

Specialization further accelerated innovation. Coming just before dawn on the last day of our calendar, the plow created a surplus of food, allowing us to specialize. When everyone needs to hunt and gather in order to eat, then crafting pots, knives, and other tools can only be done in spare time. But agriculture created a surplus, allowing some to exchange part of that surplus for technology made by specialists. Those specialists had time to create all sorts of new technologies, including those making agriculture more efficient. This creates yet more surplus and opportunity to develop even better technology. Feeding on itself, this process has led to the point that farmers, once representing near 100% of every society, now represent just a tiny percentage of the developed world—2.5% of the U.S. population.

Both *plan* and *accident* have always fostered innovation. We have stumbled across inventions (probably the first stone tools) and pursued them (the atomic bomb during World War II). It is, of course, hardly profound to say that technology is either intentional or it is not. More interesting are the characteristics of the environments that contribute to one rather than the other. Some environments focus resources on planned development. For example, war led to both the atomic bomb and the modern pencil "lead." Other environments allow a diversity of efforts, which can result in more accidental discoveries. The laser is an example. Developed in a peacetime free market, it was invented to create very high frequency radio waves. No one anticipated using it to read music from compact discs or reshape corneas for better sight.

If you spend resources on inventing a new technology, you might prefer that someone else not steal your idea. If the threat of losing out on the benefits you feel you deserve is strong enough, you might not go to the trouble of inventing in the first place. *Protection* through patents and similar government intervention are intended to encourage innovation by assigning rights to the information behind inventions.

Two factors now make the protection of information increasingly important to the creation of technology. First, information travels ever more quickly around our world and, second, ever more technology is composed mostly of information. The value of both computer software and engineered drugs is less their raw materials (e.g. a compact disc or pill) than their design (the development of which constitutes most of the technology's cost). Awareness of this pattern is important in unraveling the debates about unauthorized music sharing and about the costs of AIDS drugs in the developing world.

Understanding where technology comes from is part of our quest to understand and evaluate technology. This chapter's question is one of the four blocks in the foundation of Identity (along with what is it, why we use it, and how it works). Part of being able to understand and evaluate technology is the ability to encourage innovation in our organizations or communities. The patterns we discover in where technology comes from are not passing fads. In fact the first one has been in play for millions of years.

> Thirty thousand years ago,
> chipping flint was the
> high technology of the day.
> – Eric Drexler

Other Technology

Apes and birds use sticks as tools. Chimpanzees throw stones and use them to crack open nuts. But there is a crucial difference between the stone tools of our ancestors and sticks or simple stones. Stone tools have intentional shape, usually a sharp edge. With sharp edges, stone tools can sharpen sticks into spears, slice animal hides into clothing, or cut animal sinew with which to sew those hides together. Stone tools can create other tools, which can create still more tools, leading to a cascade of improvements.

Few stones lying around have sharp edges useful for cutting. We can create such an edge, though, by striking stones together. A chimpanzee can do this if trained, as scientists demonstrated by presenting Kanzi, a talented Bonobo chimpanzee in captivity, with a food treat enclosed in a box tied with rope. They demonstrated how to hit two stones together to flake off pieces, some of which were sharp enough to cut the rope. Kanzi learned this and even tried his own technique of throwing one stone against the other.

What Kanzi did not learn is something that we did two million years ago. Randomly smashing rocks together, which also happens in nature when they fall from cliffs or rivers pound them together, does not produce good cutting edges. To get those, you must study a stone to figure out the best angle of striking. Then, with each strike breaking off flakes, you must adjust your angle.

Few humans have that skill today, but we would have the cognitive ability to develop it if we needed it. Over months of training, Kanzi—who can do something as sophisticated as tying his shoelaces—did not develop it. What he created with human coaching was like the very first stone tools we have found. The first three steps in our development of stone tools were:

- *Stone chips* (3 to 2 million years ago) These often look similar to rocks broken by natural causes (e.g. crashed into each other by a rushing river).

- *Olduwan stone* (2 to 1.5 million years ago) They are clearly intentional because the flakes and the core pebble from which they came are found together.

- ***Handaxes** and **Levallois flakes*** (1.4 million to 250,000 years ago) Easily recognizable as tools, handaxes—what we would call an "ax head"—would later evolve into a more familiar ax when strapped to a wooden handle. Levallois flakes showed forethought because the technique involved careful preparation of the core of a rock in order to get cutting flakes of a predictable size.

Building on itself, the tool-making process blossomed to transform our world, but it is hard for us to imagine how slowly this transformation began. In the 21st century, some technology changes every few years. If a computer comes with half the memory you want, just wait two years for the next model. It will have double the memory (and processor speed and disk size) at the same price. Or, get the old model at near half the price.

We are becoming accustomed to technology flowing like a swift river, with improvements arriving constantly. Waiting thousands of years for a small improvement in technology would seem, to us, forever.

Before stone tools existed, they must have been very difficult to imagine. Further, they were not trivial to make. A trained chimp, capable of much else, apparently lacks the hand coordination and mental forethought to chip stones into the specialized tools made by our ancestors. But once the tool-making process started, it built on itself. Tools made it easier, both physically and mentally, to create better tools.

Physically, the tools gave us capability, such as sharpening sticks, which our bare hands did not provide. Mentally, the tool showed us by analogy what might be created. The more examples of tools we saw, the more likely we might think up another. Today, invention often comes of imagining something "a little like this thing and a little like that." Seeing that music or other audio can be recorded on cassette tapes and CDs suggests that video should not be restricted to VHS tapes. Hence, the DVD (which looks just like a CD).

This mental and physical effect is in play today. Computers extend our brain as stone tools extended our brawn, so we may be able to apply our insights from ancient history to the present. Physically, new computers allow us to design tens of millions of transistors onto a chip of silicon the size of a postage stamp. These chips will power the next generation of computers, allowing design of even more complex circuits.

Mentally, computers have become our starting point for wondering, "what's next?" Computers model a whole new form of tool, one that operates on information, follows rules, and even appears to think. When the computer Deep Blue beat world chess master Gary Kasporov in 1997, it was applying rules, millions each second. Manipulating huge amounts of information is becoming commonplace, and people learning about their environment today will assume that capability as a basic building block toward new technology.

If advances in technology came about, in part, because stone tools were seen as basic building blocks, then we may, by analogy, expect fantastic technologies in the future from those who now view *computers* as basic building blocks. It is hard to imagine what will be created when the most fantastic technology we have today is taken as a given, just a starting line.

While it is hard for us to imagine being ignorant of basic tools, it is easy to imagine being ignorant of missiles, acids, and bombs. How long would it take us to figure them out if we were left on our own?

Seeding the Crystal

How could developing stone tools have been so slow? Were humans that much less intelligent back then? In answer to that question, consider the island of Tasmania and an old TV show, *MacGyver*. As anywhere, Tasmania suffered from occasional famine. Unusual, however, was that for about 4000 years Tasmanians, surrounded by rich oceans, did not fish. They did not think of fish as "food," much as most Americans and Europeans rarely think of insects as food, even though people in many parts of the world recognize them as highly nutritious.

Now, on to MacGyver, the television show about a resourceful hero who uses a paper clip to short-out a nuclear missile, a chocolate bar to plug an acid leak, and a cold capsule to trigger a homemade bomb. What did he have that most people with easy access to paper clips, chocolate bars, and cold capsules lack—other than life-threatening situations every week? Information. Most of us lack information about missiles, acid, and bombs just as the Tasmanians lacked information about fish. Similarly, before we invented stone tools, we lacked information about stone tools—a seed for technology.

Our vast interrelated network of technology is like a crystal. The molecular components of crystals can float around in liquid (non-crystalline) form until they come in contact with a "seed" crystal. This seed is literally a few molecules that have already been stacked into a crystal structure. These cause more molecules to come out of solution, adding themselves to the structure. Integrated circuits are made from a giant silicon crystal grown from a tiny silicon seed. Stone tools may have been the conceptual seed from which all technology since has grown.

The key to the next source of technology is that we are not left on our own. We share ideas and build on each other's ideas. The more contact we have with each other, the more likely one person's idea will trigger someone else's improvement. One effect of stone tools—which helped us survive by helping us hunt and by protecting us from predators and the elements—was population growth. And that led to denser populations.

> More and denser population
> means more advanced technology.
> – Robert Wright

Dense populations

One good technology leads to another. Even Thomas Edison, one of the most prolific inventors of all time, built on the inventions of others. The first incandescent light bulb was invented in 1802; Edison was born in 1847; he invented the first practical incandescent light bulb in 1879. His bulbs did not quickly burn out for two primary reasons. First, he tested variations on the earlier filaments, finding some that resisted burning. And, second, he used the latest and best vacuum pumps to evacuate the air that would have allowed the filaments to burn.

But what if Edison had been isolated from the many earlier bulbs and from advances in vacuum technology? What would he have invented? And even if he had been brilliant enough to invent these precursors and then proceed to his light bulb, what about the technologies on which those precursors were built? As we just saw, those go all the way back to stone tools. Fortunately for Edison and those of us who enjoy something brighter than candlelight, he lived in an age when dense populations transmitted knowledge of inventions far and wide.

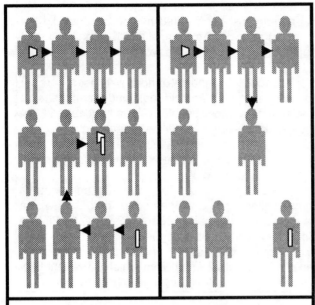

Dense population (left) is more likely to connect inventors (e.g., hand-axe + handle). Less dense population (right) is more likely to leave inventions stranded, unincorporated into more advanced technology.

For a simpler illustration consider the hand-ax, which is a sharpened stone cradled in the hand. These have been around for about 400,000 years. Someone—or, actually, many different, isolated people—invented them. Eventually, others figured out that attaching them, to a handle of wood or bone protected your fingers from being smashed and provided a lever for applying greater force. The axes we buy in a hardware store all have handles.

Did the same person who invented the hand-ax also think to attach a handle? No, not unless he or she lived for 380,000 years: the technology of hafting, or slotting a handle to cradle a stone ax head, is just 20,000 years old. Some people invented hand-axes; others invented hafting. The denser the population, the easier knowledge of one invention could be communicated to the potential inventor of the next.

When populations are sparse, ideas don't bump up against each other as readily. When Europeans discovered Australia in the 18th century, they found the native aborigines living as pre-agricultural hunter-gatherers. But southeast of Australia on the island of Tasmania, native technology was even more primitive, comparable to what Europeans had in the Stone Age more than ten thousand years ago. While Australians had at least hooks, nets, and the ability to both sew and start a fire, the Tasmanians lacked even these. Tasmanians also lacked bone tools, something developed 90,000 years earlier in Zaire (as harpoons) and 40,000 years earlier nearly everywhere else.

Why did the indigenous peoples of Australia and Tasmania not develop technology common in other parts of the world? One answer is that they lacked the dense populations that would have allowed easier sharing of information. Australia had about 300,000 people and Tasmania about 4000. With a smaller "communal memory," it is more difficult to connect two seemingly unrelated facts into a new technology.

It is even possible to forget technology. Archeologists have found bone tools, needles, and tools for fishing from about 3500 years ago on Tasmania, but none more recently. In his book *Guns, Germs, and Steel*, Jared Diamond presents the theory that technology once used on Tasmania was lost.

Chain Reaction

An analogy from physics is a nuclear reaction. In a power plant, the uranium atoms, which constitute the fuel, emit high-speed particles, which strike other uranium atoms, causing them to emit more particles. Inserting rods, which simply absorb the high-speed particles, preventing them from triggering further emissions, controls this chain reaction. Translate high-speed particles to new inventions and rods to geographic isolation.

In Tasmania, you might have been a genius, but if the other 3999 Tasmanians were not clever in just the right way, your invention would progress only as far as you pushed it. No chain reaction. Unlike high-speed particles, technology can have long lives. As long as enough people value a technology, it can be replicated by succeeding generations, waiting for the next clever inventor. But if even a single generation loses interest or the ability to fabricate it, it can be lost, just like a high-speed particle shooting off into space.

Population grew, in part, because spears, sewn clothing, axes, and other technologies protected us from claws and cold. In his book *Non-Zero: The Logic of Human Destiny*, Robert Wright theorizes that population density in Africa and Eurasia began to increase about 40,000 years ago because the growing population ran out of empty, habitable land, and was forced to simply pack in closer. Around this cusp, the rate of technological change increased from one major innovation every 20,000 years to one every 1400. Another cusp was reached about 12,000 years ago, about when the harvest sickle and fired clay pottery were invented. With proto-agriculture and the means to store surplus, the rate increased to one every 200 years.

In California's Silicon Valley, Massachusetts' Route 128, and other geographic concentrations of technology companies, we find a similar relationship. The density of talent allows for ideas to trigger other ideas. Employees jump from one company to another, cross-pollinating as they go. Nearby universities provide basic research and even more ideas. Companies merge and combine technology, paying a premium to be located in these centers of innovation with their density of ideas and talent.

In this section, we care about dense population only because it facilitates communication. But technology now does that independent of geography. The printing press shared ideas across Europe and then beyond. Radio and television made this one-way communication far faster, and telephones and the Internet made it two-way. New technology is making our interactions even more "just like being" there. And so more innovation will be spurred by virtually—not geographically—dense populations.

Again and again, people with access to
the prerequisites for food production,
and with a location favoring diffusion
of technology from elsewhere,
replaced peoples lacking these advantages.
– Jared Diamond

Specialization

More than 10,000 years ago, the advent of agriculture became the most important creator of dense populations. Ten to 100 times as many farmers than hunter-gatherers can survive in a given area, but agriculture had a much more interesting effect on human society than simply allowing it to become denser. It gave birth to the specialist, who often earned his or her food by developing and creating technology.

The last Ice Age peaked about 18,000 years ago, and its waning forced us to adapt to a new environment. Rising oceans reclaimed vast coastal areas that the earlier Ice Age had exposed (one estimate submerges 40% of all the land that was dry during the Ice Age). Jungles heated up and dried out. Grasslands replaced the dense forests that had been full of animals to hunt. The area of southwest Asia we know as the "Fertile Crescent" looked anything but fertile to our hunter-gatherer ancestors. Grasses were not food for humans…and then they were, after two things happened completely beyond the understanding of anyone at that time.

First, two plants crossed genetically, creating a mutation. The fertile combination of wild wheat and a natural goat grass provided a grain sufficiently plump to be worth harvesting. Our ancestors supplemented their diet with it and, eventually, must have learned how to plant it, too, because they were ready to exploit the second surprise.

This surprise—a second mutation—arrived about 2000 years after the first one. It made the wheat even plumper and more attractive, but so tightly packed that it was no longer able to re-seed itself with wind alone. If humans had not already learned how to sow seeds, the new mutation would have died out. But it had something we wanted: a big, nutritious seed. And we had something it needed: the technique of spreading and planting its seed. By helping it survive, we changed it. By breaking us from the often-nomadic life of hunting and gathering, the wheat changed us.

Around 10,500 years ago in southwest Asia, our harvesting of wild cereal grains became what we might call agriculture. Less than a thousand years later it developed in China. In the New World, it played a lesser role, probably because fewer native crops were as attractive for cultivation (corn and little barley, but no wheat or rice, which were key in southwest Asia and China). The symbiotic relationship that is agriculture changed technology, which spread to and affected neighbors.

We had used our sickles, baskets, and grinding tools to plant, harvest, and process wild cereal grains, but that genetically altered wheat led to countless more tools because it allowed us to produce surplus. If you have a surplus, you can trade. You need to store it, probably in the clay pots invented 12,000 years ago. How do you keep track of who produced which food and who is trading what with whom? Our answer was with writing, at first carved into clay, later on papyrus and then paper.

Since you want to protect the surplus, you may need soldiers with weapons. The surplus also fed a government that coordinated everything and craftspeople that developed improved tools for planting (e.g. the plow), harvesting, storing, fighting, and worshipping (which left the biggest artifacts, such as pyramids). The greater the surplus, the more people could specialize in something other than farming...so long as that activity produced something that people wanted. Specialization focused us on improving our tools.

Without surplus and specialization, improving tools is done in spare time. However, if your entire job is to create tools, you are much more likely to figure out how to improve them. Then, the improvements can start building on each other, accelerating technological progress. Ancient Egypt built these technologies atop agriculture:

Irrigation canals:	Agriculture. Water trapped in reservoirs during the annual flood was distributed by canal to the fields during the ensuing drought.
Plow:	Agriculture. Improved prior methods of opening the land by hand or sharp stick in order to plant crop seeds.
Calendar:	Agriculture. Prediction of the seasons was critical for knowing when to plant seeds. Observing annual patterns in weather gave the Pharaoh seemingly god-like powers of predicting the annual flood of the Nile River basin.
Wheel:	Agriculture in the form of the potter's wheel (3500 B.C.) to make containers for grain. Also used for transportation on funeral vehicles (in Sumeria), though not for building the pyramids, where skids were used for dragging the giant blocks.
Writing implements:	Communication and keeping track of food stores and transactions (writing on wet clay led to papyrus in 2500 B.C.)
Loom:	Shelter and clothing.
Cutting tools:	Building (canals and pyramids) and for making other tools.
Simple metallurgy:	Making metal tools and for religion (jewelry, decoration).

Stone tools let us create better tools, denser populations helped us share ideas, and agriculture gave us surplus, allowing us to specialize and create even more surplus, more tools, more wealth. How do we apply that wealth? What technologies result from it? At times we direct it toward a single objective. At other times we seem to all go off in different directions, and then select the best results. Both approaches are active today, with war often focusing resources and peace diversifying them.

Plowing It Back In

Agriculture made development of these technologies possible because its surplus could be reinvested. By analogy, consider a simple bank investment with interest. If the interest is plowed back in the investment grows exponentially. One hundred dollars invested with a 10% return becomes $110 after one year, $121 after two, $133.10 after three, and $259.37 after 10.

Agriculture plows its surplus back in by spending it on specialists who develop improved, more efficient technology. This creates greater surplus, which can be plowed back towards even greater technological development, and so on.

> Either history is a series of
> individual and unrepeated acts
> which bear no relation to anything
> other than their immediate and
> unique temporal environment,
> or it is a series of events triggered
> by recurring factors which manifest
> themselves as a product of
> human behavior at all times.
> – James Burke

Plan or Accident

Near the end of the 18ᵗʰ century, when war with Britain cut France off from high-quality graphite mines, French scribes could have been reduced to quills and inkwells. Government and commerce relied on scribes to write down laws, transactions, agreements, and plans. And they did this most efficiently with graphite wrapped in wood, commonly known as a "pencil." Napoleon declared the invention of an alternative pencil to be a national priority.

In 1795, Nicolas-Jacques Conté, combined readily available lesser-quality graphite with clay to produce a superior pencil "lead." Not only did this save French wartime administration and commerce, but this new pencil lead proved even better for writing than pure graphite because varying the proportion of the two ingredients controlled its hardness. For the first time, artists could have their soft pencils and architects their hard ones.

Finding a good alternative to pure graphite was quite intentional, but the benefit of varying lead hardness was an accident. Much technology comes accidentally. Earlier in this chapter we mentioned the laser and its unexpected application to reading CDs and eye surgery.

Plan and accident also work in combination. The transistor was a planned improvement on the vacuum tube. But the inventors of the transistor based their work on the germanium crystal rectifier, which was discovered entirely by accident decades earlier. The first radios were "crystal" radios, based on the germanium crystal rectifier, and in use long before the first vacuum tube or transistor (something we touched on in Chapter 1).

Another example: by chance, Alexander Fleming discovered penicillin, the first antibiotic, in 1928. Production was minuscule until World War II, when thousands wounded in battle were dying of infections that could be cured by it. By plan, production then became a factory operation: 28 pounds were produced in 1943 but seven tons were in 1945.

In a story that returns us to Napoleon, plan led to accident. In his continuing international adventures, he needed better food supplies for his far-ranging troops, so he announced a national prize for a solution. Nicholas Appert claimed it in 1810 by stuffing empty champagne bottles with fruits, vegetables, milk, and meat before sealing and cooking them. The heat killed bacteria and "bottled" food fed French armies and navies.

Through several coincidental meetings and relationships, bottled food led to Bryan Donkin and several other British entrepreneurs, who replaced the glass champagne bottles with tin cans. Nearly two centuries later, canned food is still common.

Technology comes from both plan and accident. Plan may focus resources on goals, but, before we've invented something, we often don't know just how to do it or whether that something is possible and even desirable. Chance works with a diversity of explorations, so that many can fail while a few produce surprising results.

Tactics of Bacteria and the Drugs They Battle
The development of technology by plan and chance are comparable to the different approaches taken by pharmaceutical manufacturers and the infectious bacteria they develop drugs to attack. Drug development is typically focused on attacking a particular disease. In response, the bacteria have no focus or central plan.

In their diverse multitude, bacteria simply mutate and trade genes with other bacteria. Natural selection culls out the trillions of unfit and rewards the fitter millions. Because the bacteriological strains resistant to our treatment no longer have to compete with their relatives we killed, they can reproduce into their original trillions. And so medicine goes back to the laboratory to develop a new weapon. Evolution led to the intelligence that focuses resources, and it continues to operate by pitting rapidly mutating bacteria, evolving a diversity of possibilities, against that intelligence.

Several of our examples for plan have come from war, which focuses resources to support grand plans. The Cold War contained the Space Race, which focused Soviet and American resources as a conventional war would have. It kicked off in 1957 with the Soviet Union launching Sputnik, the first earth-orbiting satellite, and started to wind down in 1969 when the U.S. landed men on the moon.

> At base, the momentum for the arms race
> is undoubtedly fueled by the technicians
> in government laboratories and in the
> industries which produce the armaments.
> — Solly Zuckerman

Some suspect that war is used as an excuse to develop technology. The "war on terrorism" of the early 21st century focused resources on surveillance and information technologies such as the U.S. government's "Total Information Awareness" system. It was quickly renamed the "Terrorist Information Awareness" system to assure the public that, although everyone's activities would be monitored, the system would not be used against the innocent. The war on terrorism is also leading to renewed development of nuclear weapons—this time for routing terrorists out of deep bunkers or mountain caves—and for a missile defense shield.

While plan may result in focus, and hence efficiency, a diversity of efforts gives us the most chances at valuable accidents. Silicon Valley, in California, is known for thousands of start-up companies, pursuing wildly varying ideas—often failing, but sometimes succeeding. The successful companies are lauded, inspiring others to copy them. The failures make for even more entertaining discussion and reading, helping others to learn from those companies' mistakes. Once someone tries something, others can observe whether it should be copied, improved upon, or avoided.

For example, before developing its nuclear power industry France observed what the U.S. had accomplished by "accident." A diversity of competing power companies in the U.S. developed many different nuclear plant designs. The cost to certify and build these often-custom jobs was high, limiting their appeal. The French government focused the approach, mass-producing just three types of power plants, and provided financing. Now, nuclear plants supply approximately 75% of French electricity, but little more than 20% of U.S.

A relatively free market in which a diversity of people, organizations, or countries can pursue their own aims makes for many fruitful accidents. Some opportunities may require focus of resources, so diversity and focus complement each other to develop technology out of accident and plan. In our lives, we will observe environments of focus and of diversity. History shows us how they differ as sources of technology. Because our decisions influence which environments will emerge, it benefits us to know when to choose the efficiency of focus and when to choose the resilience of diversity.

> Congress shall have power…
> to promote the progress of science
> and useful arts, by securing for
> limited times to authors and inventors
> the exclusive right to their respective
> writings and discoveries.
> – Article 1, Section 8 of the U.S. Constitution

Protection

You spend much of your life pursuing a dream of a better mousetrap. Through research, trial, and error, you finally devise a design that makes all existing mousetraps appear primitive by comparison. You show your design to those who might finance its manufacture, but to your horror, they make and sell the new mousetraps without your control and without sharing the profits. It would be enough to make inventors give it all up and leave the world to get by without technological improvements, leaving the "pirates" with nothing to pirate.

Recognizing how fragile the investment in developing new technology can be, Venice created a patent system in 1474 and England did so in 1624. Just as the patent system in the U.S. today, these granted a temporary monopoly on creation of a specified invention.

In theory, this encourages more development by protecting the investment. And there is a further benefit of publicly documenting an invention. It allows other innovators to learn from the design and make improvements significant enough that they warrant their own patents. Either way, society reaps the reward of increased technology development. In practice, patents can also inhibit development of technology. A broad patent protects so much that many potential inventions would infringe upon it, discouraging anyone other than the patent holder from developing them (see sidebar *Yertle the Turtle*).

The incandescent light bulb that Thomas Edison patented in 1879 shows the value of patents in promoting innovation. Edison tested 3000 different filaments before finding that carbonized cotton thread did not quickly burn out. With a patent for the design he developed, he produced light bulbs.

Patents are crucial for technology with high information content. The development of a new drug costs about a quarter of a billion dollars (including a share of the developments that failed), but very little to manufacture. Without patent protection, no one would dare invest because a competitor could simply copy the research to avoid most of the cost and all of the risk. Patents give some assurance that pharmaceutical companies will be able to charge much more for each pill than it costs them to make it, so they can reward the investment that went into its development and fund further development.

And, by and large, the patent system has worked. Over the past few centuries, those areas with protection of intellectual property have been major sources of technology. But to ensure that it does not stifle innovation, technology protection must evolve with technology.

Yertle the Turtle

In Dr. Seuss' children's book *Yertle the Turtle*, a turtle with delusions of grandeur believes that he is owner of all he can see. As the dominant turtle, he compels the other turtles of the pond to stack themselves skyward to afford him a higher throne, broader view, and greater kingdom. The moral comes when Yertle's reach exceeds his grasp, his tower becomes unstable, and he comes tumbling down into the pond. Covered with the muck from the bottom of the pond, discovers that he can see nothing else.

While Yertle had been able to maintain his claim to the pond by whatever means a dominant turtle remains dominant, with respect to the many lands beyond the pond, he confused *seeing* with *owning*. Similarly, protection of intellectual property can confuse an event (identifying) with a process (developing). Technology comes from an environment that protects its development, not merely its identification.

Patent attorney Dennis Fernandez came up with the idea of television viewers seeing each other and discussing a shared program. Fernandez has a U.S. patent for televisions with cameras that would show both a program and a headshot of another viewer, who could be far away. Although he is also an electrical engineer, he has no plans to actually make one or more of these devices. He is satisfied owning the intellectual property rights so that anyone who does try to make one would have to pay him. This patent protects someone who identifies something and discourages anyone who would develop it.

Two extreme cases of "Yertle the Turtle" patents are for the wheel (U.S. 5,707,114) and the genetic code for a person living in a remote area of Papua New Guinea (U.S. 5,397,696). These are not "rewards" for investing in the development of new technology, but the equivalent of squatting on public land and claiming it as ones own.

In biotechnology, gene sequences have been covered under patents, even though they are not inventions and their function may not yet be known. Some scientists complain that this departs from the intent of patents to promote innovation because it allows someone to "squat" on a gene sequence without developing scientific knowledge about its function. This discourages others from developing that knowledge.

Without a commitment to develop, those who lay claim to an area of information, whether the design of the wheel or the DNA of a person, are just blocking innovation. Even during the 19th century land rush in the U.S., when the government awarded 160 acres of land to whoever staked it out first, only those still working the land after five years received title.

Innovation occurs for many reasons including greed,
ambition, conviction, happenstance, acts of nature,
mistakes, and desperation. But one force above all
seems to facilitate the process. The easier it is
to communicate, the faster change happens.
— James Burke

Easier communication leads to faster change because information fuels innovation. In each of the sources of technology in this chapter we find the thread of information.

Other technology gives us greater physical capability to create new technology, but because it gives us examples of what is possible, suggesting combinations or analogous technologies, it also conveys information. Inquisitive minds see not only the tool, but the information presented by the tool. The tool may be useful for a different task or, since this tool does, indeed, exist, then an even better tool could also exist. The information in the tool points toward new applications and new technology

Improvements in technology were at first isolated and sporadic, but as *populations* grew *denser*, information about technology (and anything useful) reached more people. Communication cascades as the information that one person shares with ten others is passed on to a hundred more in a chain reaction of development. Information about how to create and use technology can even loop back to the original inventor, who may improve upon the improvements. Today, advances in communication technology gives us many of the same benefits of dense populations.

Specialization, made possible by surplus, leads to diverse technology, each of which is surrounded by information. Carpenter, potter, metal smith, and alchemist developed their own vocabularies in the dawn of agriculture, much as scientists, technicians, engineers, and other specialists have in our current era. We use this vocabulary to describe and design ever more sophisticated technology, which, like denser populations, is a process that feeds on itself by creating the need for more specialized information.

Both *plan* and *accident* are also about information. The "plan" of planned innovation is simply information about objectives and resources to be used in developing a technology. And accidents create information by revealing patterns in the Universe of which we were unaware—which sometimes lead to new technology. Some environments are particularly suited to creating information through accidents (e.g. peacetime free market) and then focusing resources to exploit that information (e.g. war). Penicillin was accidentally discovered in peacetime and then, by plan, manufactured in useful quantities during World War II.

Patent *protection* gives property rights over information about an invention and publicizes that information. This makes the investment in innovation more attractive and can give other innovators new ideas. As information becomes a greater component of technology (e.g. software and engineered drugs), technology becomes easier to copy and protecting it becomes that much more important. But as information travels more quickly and easily around the world, protecting it becomes more difficult. And protection can also inhibit innovation by assigning broad rights for discoveries rather than developments.

There are countless sources of technology, provided one is willing to delve deeply enough into the details. But for our purposes, these answers are specific enough. In the previous chapter on why we use technology, we made the point that our intent is not to create an exhaustive list but rather to clarify the question, to give it context. For in our quest to understand and evaluate technology, it is the questions that have enduring value since new times and new technologies will bring new answers. The answers in this chapter are a starting point, not an encyclopedia from which all answers will come.

One benefit of being able to understand and evaluate technology is that we know how to promote development of technology. We may find ourselves in position to influence our environment, perhaps by voting on a law, leading an organization, or developing a school curriculum. The patterns in this chapter, drawn from the millions of years we have had technology, can inform the decisions we make.

Chapter 4

How does Technology Work?

Hidden beneath the surface,
technology of all descriptions
works according to a few,
simple principles.

Early stone tools took skill to create. Even with many hours of practice, modern anthropologists cannot match the skill Cro Magnon and Neanderthal had for chipping away at rocks to create sharp cutting edges. Each strike of stone against stone chips away rock and changes the optimal angle of the next blow. Of course, tens of thousands of years ago, we spent years learning how to make knives from stone because it was a matter of survival.

But as difficult as the technique may have been, understanding how the technology worked was simple. It had no moving parts. Little by little that has all changed. Modern technology works in ways so complex that only teams of experts can understand them completely.

Take a car, for instance. Specifically, consider the brakes. Mechanical engineers understand the physical interaction of brake shoe and rotor or drum. For antilock brakes, we need a computer engineer because computers monitor each tire for impending skid. But few computer engineers understand all the layers within computers (software engineers may not understand the hardware and hardware engineers often divide their field into designing digital logic circuits, analog circuits, microchips, and more), so we would need a team of engineers to explain exactly how the brakes on a modern car work.

Even immortality would not allow an individual to grasp all of technology's workings because specialists are inventing anew faster than anyone can keep up. Of course, we can console ourselves that we don't *need* to know how every technology works. The beauty of specialization, made possible by agricultural surpluses, is that we can (and do) delegate tasks to specialists, such as engineers, technicians, and scientists.

But delegation is different from abdication, which is what we do when—knowing nothing of technology—we let the specialists make all the decisions. We may have to rely on experts to process the details, but we can equip ourselves to comprehend their analyses, opinions, and predictions. Threading through a sea of technical details, are simple patterns that connect wide varieties of technology, explaining aspects of how those technologies work in common sense terms.

There is a pleasure in discovering simple patterns just behind the apparent complexity of technology—and seeing how they have remained true over time. In this chapter we examine seven that pop up in the functioning of a variety of technologies:

1. All technologies rely on *energy*, from a horse pulling a plow to gasoline fueling a car. Many convert energy from one form to another.

2. Technology can be ***distributed*** into many small parts (e.g. power generation at home with solar or wind) or ***centralized*** (e.g. nuclear power plants or hydroelectric dams), sometimes alternating between the two as new inventions make one better than the other.
3. Bicycles, nuclear power plants, airplanes, and missiles rely on feedback and correction, two key elements of ***control*** systems, which keep much technology focused on the goals we set.
4. ***Information*** is the difference between a compact disc that comes in junk mail and one containing the human genome. In the form of rules for solving a problem, it is an ***algorithm***, which enables us to understand a technology's behavior without having to understand its implementation.
5. ***Repetition*** and ***layers*** are two ways that complex technology can be composed of simple building blocks (much as the repetition of 26 letters and the layering of words, sentences, paragraphs, and chapters compose this book).
6. ***Emergent behavior*** is about the whole being more than the sum of the parts. Just as an ant colony behaves very differently from any individual ant, so, too, do many complex technological systems behave differently from any of their components.

Since these patterns have endured over time, we may find them threading through future technology, however strange and foreign it may appear. We still need experts to design, build, maintain, and explain. But, unless we choose to cede control of our future to those experts, we need a basis from which to understand their explanations and to form our own evaluations. That basis starts here.

Energy: the muscle behind technology

Here's a trick that won't work. Wire a solar cell (which generates electricity when exposed to light) to a light bulb that shines on it. This system creates its own energy, running the light from its own light. A similar trick was proposed in the waterwheel invention of Robert Fludd, a 17th century London doctor. In theory, water flowing over the wheel powered a pump that sent all the water back upstream so it could again turn the wheel. If you could tap a bit of the energy from the solar cell or Fludd's waterwheel to do work—perhaps run a stereo or grind wheat—then you'd perform work for free: no external source of energy necessary.

These two contraptions do not—and cannot—work. Commonly referred to as perpetual motion machines, the U.S. Patent Office will not issue patents to them, though inventors still try. The flaw in the solar cell and bulb system—just as in any perpetual motion machine—is that some energy is always lost: By the numbers:

15%	of light striking solar cell is converted into electricity (85% reflects or becomes heat)
x 99%	of the electricity going through wires is not dissipated as heat
x 50%	of the electricity flowing through the light bulb becomes light (50% becomes heat)
x 25%	of that light may actually strike the solar cell (75% light shines elsewhere)
= 2%	of the light that strikes the solar cell would become light striking the solar cell again

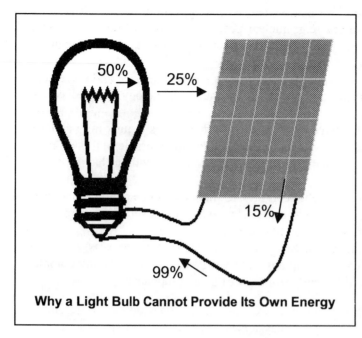

Why a Light Bulb Cannot Provide Its Own Energy

Any energy we might initially endow the system with would quickly disperse in the form of heat and of light shining somewhere other than onto the solar cell. And even if every part of the system could be 100% efficient, tapping any energy to do work would quickly exhaust the system's internal supply. Without a continuous external source of energy, our contraption would just sit in the dark.

Technology—just as everything else in this Universe—needs energy to do work. So, no matter how complicated and confusing a technology may be, here is something we do know about it: somewhere, there is a source of energy.

Of course, energy sources can be inconspicuous. For example, most wristwatches can operate for years on a tiny battery hidden inside, but eventually the chemical energy stored in the battery is exhausted and the watch stops. Some watches even derive energy from the motion of the wearer's arm or the heat differential between the arm and ambient air. No battery, but still an external source of energy fueled by the food eaten by the person wearing the watch…and not the magic of perpetual motion.

Food was the source of energy for the first technology humans developed, and is still the only source for the tools animals use. Energy in food is transformed to energy in muscles, which operate sticks, stones, hammers, scalpels, bicycles, and other manual technology. The energy in food comes from nuclear fusion in the sun, which creates light, which creates plants through photosynthesis, which feed animals, which feed other animals.

Though dogs may have started living with humans 14,000 years ago, they were not the first animals we tapped to power technology. Domesticated reindeer started pulling sledges in Northern Europe about 7000 years ago, 1000 years before we domesticated horses and 3000 years before we got them to pull vehicles. But even with domestication of animals, the only energy source for technology was muscle powered by food.

That all changed with the sails on ships, which, more than 5500 years ago, harnessed wind. Sails are limited to propulsion, but waterwheels, invented about 2100 years ago, used moving water to grind wheat, corn, and sugar cane, work bellows to make fires hot, and pound hammers onto rocks and metals. Windmills, first created around 1600 years ago, were applied to many of the same tasks. The sun, again, is the source of these energies, whether by differentially heating the atmosphere to create wind or by evaporating water to create rain and flowing water.

Fire, long useful for cooking food and staying warm, first became an energy source for technology around 100 AD. Greek temples used steam turbines to open and close doors, as if willed by the gods. Whether limited by materials, imagination, or motivation (a huge slave population already provided manual labor), the Greeks did not develop the steam turbine into a practical energy source. Over time, the steam turbine was forgotten, but, by the year 1700, it was reinvented and went beyond novelty tricks to power steamboats, steam trains, and—for a short time—steam cars, too. Burning wood releases the energy captured from sunlight through photosynthesis.

So does burning fossil fuel. Coal, oil, diesel, and gasoline were plants, dinosaurs, and other animals, before millions of years of high-heat, high-pressure subterranean processing. In 1860 the internal combustion engine tapped fossil fuel's explosive energy. Now the burning of diesel and gasoline propels most of our transportation and the burning of coal provides more than half of U.S. electricity. While not renewable, fossil fuels originally got their energy from the same source many renewables do: sunlight and photosynthesis.

Propelling arrows with gunpowder, 13th century China tapped chemical energy. While gunpowder has been wildly successful in creating explosions for guns and mining, several attempts at using it in internal combustion engines failed. Another source of chemical energy has proven an excellent source of energy for technology. In 1800, the invention of electrical batteries allowed us to convert the chemical potential in various metals into electric current. Some evidence suggests that two millennia ago chemical batteries were used in Iraq, but it remains speculation. Reversible chemical processes are used in rechargeable batteries, especially useful in laptop computers and cellular phones.

In 1943 we harnessed nuclear energy. Nuclear fission (splitting large radioactive atoms into smaller ones) is used as both an energy source (nuclear plants) and a weapon (atomic bomb). The sun, too, uses nuclear energy, but in a fusion process (combining small atoms into larger). The only Earth-based fusion reactions producing significant energy have been uncontrolled, and so used only as weapons: hydrogen bombs. Although we have yet to control a fusion reaction to produce energy, for billions of years the fusion reaction located a safe 93,000,000 miles away from Earth has been source to most of the forms of energy we discuss in this section.

Another source not leading right back to the sun is geothermal energy. Heat deep within the earth from both gravitational pressure and radioactivity creates a temperature differential, which can be harnessed to produce energy. The classic view of geothermal energy is of steam shooting from the ground. Like steam from other sources (e.g. wood or coal fires or nuclear fission reactions), this can drive turbines, which generate electricity.

Back to the sun. In 1954, 115 years after the principle of converting sunlight directly into electricity was discovered, photovoltaic cells, or "solar cells," made it practical. The sun's light causes electrons to move, which is electricity. Solar cells are important sources of electricity on earth-orbiting satellites, the International Space Station, handheld calculators, and some buildings and homes. Centralized generating plants are few and small compared to fossil fuel or nuclear plants.

How Solar Cells Work

Solar cells are semiconductors, similar to integrated circuits. These silicon products are descendants of the germanium rectifier that made possible 19th century "crystal" radios (which we touched on in the chapter **What is Technology?**) A rectifier allows electricity, the movement of electrons, to flow in only one direction. This is crucial in solar cells because when a photon of light liberates an electron in the absence of a rectifier, that electron can fall right back into place, releasing the energy it absorbed from the photon as another photon. Light goes into the material and light comes out.

But if a liberated electron is caught on the wrong side of a rectifier, it cannot return to the hole it left. If the easiest way around the rectifier and back to the hole it left is through wires, light bulbs, motors, or televisions, then that electron will go that way, driving our appliances. The more photons, the greater the imbalance of electrons on one side of the rectifier and holes on the other, the more force with which those electrons will flow through an electrical appliance to restore the balance.

Almost all the electricity we consume comes, not from these semiconductors, but from moving magnetic fields. It's a physical law that electrons move in a conductor (e.g. wire) when a magnetic field moves nearby. The turbines spun by moving water in hydroelectric dams or by steam in coal, oil, or nuclear plants spin magnets near coils of wire, generating electricity. In the year 2001, the U.S. generated 494 million kilowatt hours from solar, representing just 0.013% of the electricity used that year. Even some of the solar power came, not from solar cells, but from moving magnetic fields, with sunlight heating water into steam to drive turbines.

In the use of fossil fuels we see some remarkable transformations of energy. In the sun's fusion reaction, matter is converted to energy (according to Einstein's famous $e=mc^2$ equation) at the rate of 4,600,000 tons per second. Of the energy released from the sun, only one part in two billion reaches the earth. Millions of years ago, some of that was consumed by photosynthesis in plants to create sugar molecules. These molecules fueled growth of those plants, some of which were consumed by animals. On dying, some of these plants and animals did not decay. Instead, deprived of oxygen, they began a subterranean process involving pressure and heat that, over millions of years, produced fossil fuels. These carbon compounds, dense with potential energy, combust to produce heat to produce motion.

In combustion engines fossil fuels drives wheels of vehicles. A 2003 study calculated that 16 acres of wheat (and millions of years) would be required to make a gallon of gasoline, about enough to travel 20 to 40 miles in a typical car. In power plants fossil fuels produce electricity. The process of generating electricity from sunlight by way of coal, can be broken into 11 steps:

1. Sun shines on ancient plants, fueling photosynthesis
2. Photosynthesis creates sugar molecules, building blocks for biological growth
3. Plants and animals die but, deprived of oxygen, are prevented from decaying
4. Remains sink deep into the earth to be heated and crushed for millions of years
5. Miners dig coal from the earth
6. Trains transport coal to power plants
7. Coal is crushed and refined
8. Coal dust is combusted to heat water

9. Steam expands and pushes turbine
10. Turbine spins with magnets attached
11. Moving magnetic field induces electric current in nearby coils of wire

From there, electricity drives technology, which continues to transform energy. Light bulbs turn electricity to light. Motors convert it into motion (which in a refrigerator drives a pump that creates a temperature differential). Heaters and electric ovens convert it to heat. Microwave ovens convert it to microwaves, which interact with foods and beverages to create heat. Televisions and stereos convert it into information-laden light and sound (another form of energy).

The history of technology includes many of these energy transformations. The first diesel train locomotives, like the steam locomotives that they replaced, had linkages to each of the drive wheels. Success of the diesel locomotive came with a counter-intuitive idea: use the diesel engine to generate electricity, which is wired to an electric motor on each of the drive wheels. This sequence of fossil fuel to mechanical energy to electricity and back to mechanical energy eliminated heavy and inefficient mechanical linkages or transmissions. This back-and-forth conversion is so efficient and economical that the diesel-electric engine rapidly replaced the pure diesel, and is commonly referred to simply as the "diesel engine."

Submarines benefited from the same diesel-electric combination. Until the snorkel, invented by a Dutch officer in 1933, submarines could not use their diesel engines while submerged because combustion requires large quantities of oxygen. The snorkel let them draw in oxygen while running just below the surface, hidden from radar. However, diving to safety deep below the surface still meant running on electricity. Bulky transmissions to connect either the diesel engine or electric motor to the propeller were eliminated by connecting the diesel engine to a generator, which both recharged the batteries and powered the electric motor now connected directly to the propeller.

This approach persists in nuclear submarines, whose generators derive heat from a fission reaction, which does not need oxygen (nor does it produce exhaust, which would reveal it's submerged location to hunting ships). The heat turns water into steam to spin a turbine that turns a generator that produces electricity. As in the diesel-electric submarines, electric motors turn the propeller. Unlike the diesel-electric submarine, the nuclear submarine can run completely submerged until food runs out, since it produces both oxygen and drinking water from seawater. Land-based nuclear power plants take an approach similar to the submarines. Heat from the fission reaction turns water to steam, which spins a turbine to generate electricity.

"Fuel cells" in hydrogen vehicles demonstrate energy transformation by combining hydrogen from storage tanks with oxygen from the air to create water and electricity, which runs electric motors attached to the wheels. These "zero-emission" buses and cars release pure water as exhaust. It stands to reason that if energy is released by combining hydrogen and oxygen into water, then it must take energy to separate them back out. Otherwise, we might just have the perpetual motion machine science maintains is impossible. Various technologies are under development to perform this separation of hydrogen, but the simplest is running electricity through water. So it is possible that a zero-emission vehicle would run on electricity from hydrogen separated from water by electricity generated by burning coal. This system as a whole is, clearly, not zero-emission. But it does illustrate our main point: every technology requires energy and many technologies transform it.

When energy is as ready as switching on a light or pressing a gas pedal it is easy to ignore. Imagine being stripped of advanced technology for generating, converting, and consuming energy. Thousands of people begin foraging for sticks shortly after dawn. Before dusk, they shoulder bundles of wood for miles, selling them in towns to buy food for another day. Burning wood cooks food, heats homes, and drives primitive industry. To see this in the 21st century visit Ethiopia, where 90% of consumed energy comes from biomass: wood, charcoal, and cow dung. The capital, Addis Ababa, draws 15,000 "women fuelwood carriers," walking up to 10 miles with loads of 70 to 100 pounds, which sell for as much as 70 cents. The fuel carriers are acutely aware of where energy comes from.

That awareness would be useful for those who can casually tap hundreds of horsepower in their cars, similar amounts in their homes (for heating, lighting, appliances, and entertainment), and more in elevators, airplanes, and climate-controlled businesses. Choices, amplified by and dependent upon technology, are informed by such awareness. The amount of energy we control has increased dramatically over the millennia and it appears that the trend will continue. Far in the future, when technology may appear quite unlike anything we have today, understanding and evaluating it will still depend on grasping the source of its energy.

Organization Part 1: Centralized vs. Distributed

Another characteristic that will persist into the future is the organization of technology systems as either centralized or distributed. The movie *Back to the Future* depicted a "Mr. Fusion" reactor mounted in a car, energy production distributed to the point of use. But there are good reasons that current nuclear energy production is now centralized: nuclear power plants are complex, expensive, dangerous, and require expert maintenance. Other technologies, including solar cells and windmills, can be either centralized or distributed, with advantages for each. History shows many technologies swinging from centralized organization to distributed and back, influenced by new capabilities, concerns, or requirements.

The "monster in the basement" was what Mrs. Vanderbilt called the steam engine that ran an electrical generator to power the new electric lights in her house. The fabulously wealthy Vanderbilt family had replaced their candles and gaslights shortly after Edison's 1879 invention of the incandescent light, but before he developed a centralized power system. The monster in the basement was an example of a distributed power system and, with visions of high-pressure steam exploding through the floorboards, Mrs. Vanderbilt came close to throwing it out.

Until the 19th century, many factories were built next to rivers so that waterwheels could provide power to grind wheat, spin cloth, or cut wood. That distributed generation has become centralized, with most power today generated by large oil, coal, gas, nuclear, or hydroelectric plants.

Distributed 11th vs. Centralized 21st

In 11th century England, 5624 water mills provided most of the non-muscle power to between 1.25 to 2 million people. 1000 years later, in the 21st century United Kingdom, 177 power stations generate electricity to support almost 60,000,000 people. Another way to look at it: about 300 people were supported by each water mill in the 11th century and many more than 300,000 people by each power plant now. So there has been more than a 1000 to 1 move toward centralization.

Some of the factors we did not consider tend to offset each other. Unlike 11th century inhabitants, 21st century people use more than electricity (e.g. fossil fuel for cars, trucks, airplanes, and trains). But we can safely say that 21st century people consume more electrical energy than 11th century people consumed of any kind of energy. A millennium ago, there were no electric lights, computers, refrigerators, microwave ovens, or televisions on which to spend energy.

So, even though each 11th century inhabitant consumed less energy, they had 1000 times as many power plants per person than we do now. How can this be? Our few, centralized plants are far larger and more powerful than the many distributed plants they had.

A friend of the author grew up on a small farm in Iowa, where a windmill provided energy for lights and a radio. In the 1940s, power lines marched past, erected by the new Rural Electrification Administration (REA). The price for connection to the reliable and centralized power? Dismantling the windmill. The REA wanted no competition from that old farm. Centralized energy production was eradicating distributed, but, like a pendulum, the trend can also swing back.

In recent decades more people have started generating their own electricity with windmills in their yards or solar cells on their roofs. This distributed approach can be attractive for remote locations, which are often expensive to connect to the electrical distribution grid.

This pattern of centralized vs. distributed organization also shows up in technologies other than energy generation. Until the 1940s, residents of Key West, Florida, used cisterns to catch rainwater, a highly distributed water system. Then the U.S. Army built a central reverse-osmosis desalination plant to replace cisterns on the one-mile by four mile island off the Florida coast. By order of the U.S. Army, all cisterns were filled with salt water, covered over, or somehow disabled to prevent any diseases harbored by standing water.

Surprisingly low water pressure from the new system spurred nocturnal work habits among residents, who often washed laundry at 2:00 AM when water pressure was a little higher. Even at low pressure, the desalination process was expensive, so it was replaced with a 16-inch pipe that brought water from the mainland. Vulnerable to storms, the new pipe inspired little confidence and cisterns returned, cleverly disguised (one as a fireplace) to avoid the U.S. Army's unwanted assistance.

What factors cause technology to organize in a centralized or distributed way? One is technological capability. Because mid-20th century computers were huge, expensive, and required expert maintenance (and security and air conditioning), they were centralized. Late-20th century microprocessors were tiny, cheap, and ran without maintenance, so they were distributed (to microwave ovens, VCRs, and the doorknobs of hotel rooms)...except when other factors came into play. Some computing problems were so large (tracking every purchase at Wal-Mart) that they needed many coordinated microprocessors, so these were centralized...until the Internet and new software allowed some of these tasks to be distributed (e.g. calculation of protein folding patterns or the search for prime numbers).

Environmental conservation and volatile energy supply are two other factors that affect the organization of technology. Environmental conservation has encouraged some Californians to invest in solar power for their homes. In 2001, unpredictable energy supplies in California (prices jumped as high as fifty times previous levels and blackouts rolled around the state) prompted another movement towards decentralized solar energy production.

By contrast, convenience and simplicity of distribution have pushed energy technology toward centralization. Distributing electricity is much easier than solid fuels, such as coal. Centralized coal-fired electrical generation saves us from having coal dropped off at each of our homes, as was once done for heating purposes. If we wished to use it to generate electricity, we would each have to operate and maintain our own little power plant. This would be much harder than training a dedicated and expert staff operating in shifts around the clock to operate and maintain centralized power plants.

Efficiency is another advantage of centralization, given current energy technology. Not only are larger plants more efficient, but they can run continuously because someone is always using power. It would be a waste to run our own plant when we are not consuming electricity or to shut it down and start it up each time we do. When large plants have to shut down for maintenance, starting them back up is a lengthy and expensive process, so it happens as infrequently as possible.

But just when we are sure that the choice between centralized and distributed organization is obvious, technology can change and so can the best choice. Thomas Edison's first electrical power plant was in downtown New York City. A high density of consumers was crucial because Edison used direct current (DC), which does not travel well. Over long distances, most of its energy is lost to heat. Clearly, distributed power production appeared to be the future of electricity. And then technology changed.

Nikola Tesla invented and developed technology for alternating current, which can travel well over long distances. Today, alternating current dominates (alternating 60 times per second, or 60 Hertz in the U.S.) and towers hundreds of feet tall march across rural areas, carrying extremely high voltage from centralized power plants to far away consumers. Since then, the long-term trend has been toward centralization, as technological advances allowed larger, more efficient fossil fuel and nuclear plants.

New technology may well follow trends from centralized to distributed and back again. Having seen this pattern in energy production, water distribution, and computers, it may be easier to identify elsewhere. Centralized and distributed systems have been tested over thousands of years in technology and over millions of years in biologic systems—plants and animals. Studying the tradeoffs made and the environments in which each approach has been most successful may save us from making costly mistakes.

Control: like riding a bicycle

A completely different pattern in how technology works concerns control. What does "control" mean in technology? Bicycling explains.

Bicycles are not stable. Rushing along on just two wheels, we are constantly falling to the left or the right, but once we've mastered "balance" we subtly turn the handlebars in the direction of the fall and start the process going to the other side. Don't believe it? Try holding the handlebars dead straight or—easier to do—drop both wheels into a narrow channel just wide enough for the tires, like the rut between railroad tracks and a crossing street. Wear a helmet, gloves, and appropriate body armor—and do not try this where trains are running.

OK, for liability reasons, please do not try this at all.

To ride we need two things: *feedback* of which way we are starting to lean over and *correction* with the handlebars. This section is about how control (feedback and correction) is a pattern common to many technologies.

For most of history, control has been provided by the human component in the system, such as the rider on the bicycle. Even in the striking of one stone against another to create a sharp edge there is an element of control. After each strike, the human observes how the stone has chipped (feedback) and adjusts the next strike accordingly (correction). One of the oldest technologies to *appear* to control itself without a human in the control loop is a curious novelty from ancient China.

Living in China around 300 AD, you might have been lucky enough to see the strangest carriage in the world. A small statue pivoted on top of the carriage to point south, no matter which way the carriage turned. How did it work? The magic was performed by gearing very much like the differential in a car's transmission, which allows the left and right wheels of a car to turn at different speeds. And this is critical since as a car turns right, for instance, the left wheels have to travel farther and spin faster than the right wheels. In the case of the Chinese carriage, the different speed of the left and right wheels determined how far to rotate the statue.

Whatever direction the statue started out pointing (presumably south), the mechanism would keep it pointing... assuming that the wheels did not slip on the ground and all the machinery were perfectly precise. In practice, imperfection and slipping would be present. Small errors would accumulate and eventually, the statue could be pointing any direction, so it failed to become anything more than a novelty.

The South Pointing Carriage fails to incorporate control because it lacks feedback when it wanders from pointing south and it lacks correction to reposition it to point south. What the ancient Chinese needed was a control system that incorporated something they had already invented: the magnetic compass. And something invented over 1600 years later would have been useful for monitoring the compass and signaling for appropriate corrections: the microprocessor. Its invention in 1971 has done more to remove humans from the control loop than anything else in history.

Microprocessors can monitor sensors, follow algorithms, and operate motors, lights, and other electric devices. With modern technology, the inventors of the South Pointing Carriage could have monitored a magnetic compass with a microprocessor. They would have programmed it with an algorithm to run an electric motor to rotate the statue until pointing south.

This ability to monitor a condition, follow an algorithm, and operate devices makes microprocessors good at the feedback and correction necessary for control. So good, in fact, that they have invaded many technologies, such as thermostats, antilock brakes, microwave ovens, and video games. This is beginning to remove humans from the loop, though we are still involved in setting the parameters, such as the temperature that the thermostat should try to maintain. As technology becomes more sophisticated, requiring finer and quicker control, and computers become ever less expensive, we may find human control farther and farther removed.

In nuclear power plants, airplane cockpits, and intercontinental ballistic missile launch sites the stakes for control mistakes are high. So how should they be controlled? Humans are more fallible than technology, which shows few tendencies to sleep, forget, drink, gamble, or get into compromising situations subject to blackmail. On the other hand, humans have a much greater contextual awareness than does technology, which does not yet read the morning paper or know about circumstances that might impinge on a decision.

For now we compromise by keeping both technology and humans in the control loop of critical systems. Technology does the repetitive (check the temperature 10 times per second, 24 hours a day) and humans make the really big decisions (flood the reactor core).

A twist in history gives us an example of human control being inadequate. Mid-20th century, a new airplane design called the "flying wing" was introduced. The military aircraft was all wing and no fuselage, a very efficient shape. Unfortunately, it is also an unstable shape and human pilots had difficulty controlling it. So much difficulty that the design was abandoned.

Near the end of the 20th century, the designers of the stealth bomber independently came up with the same flying wing shape. At that time, computers could be used on-board to operate all of the aerodynamic control surfaces, keeping the plane stable and flying in the direction the pilot indicated. Like antilock brakes, the on-board computers were able to make many comparisons and corrections each second. There is an interesting coincidence between the stealth bomber and the old flying wing. The stealth designers used computer simulations to determine the optimal dimensions of the modern plane. They came up with a wingspan of 172 feet. Later, when they realized how similar their new plane and the flying wing appeared, they looked up the old specifications. The flying wing had a wingspan of 172 feet.

One way to look at this: we relinquished control of one technology (airplanes) to another (computers). Essential to the feedback and correction in a control system is information. Correction is based on feedback, which is information. Modern airplanes have replaced mechanical linkages between the pilots' controls and aerodynamic control surfaces (e.g. flaps, ailerons, elevators) with wires. Called "fly by wire" systems, it is clear that information in the form of electrical signals is at the heart of the system. The stealth bomber took this farther by letting an information-processing computer operate the controls. Information is at the heart of many technologies.

> If there were something like
> a guidebook for living creatures,
> I think the first line would read like
> a biblical commandment:
> Make thy information larger.
> — Werner Loewenstein

Information: algorithms

Information controls how technology works. It has since long before the stealth bomber. Silk looms in 18th century France stored information about the patterns to be woven as pegs on a cylinder, holes on strips of paper, and holes in cards (invented by Bouchon, Vaucanson, and Jacquard, respectively). That information controlled which colored thread of silk the loom wove through the fabric at each of thousands of steps. This automated the fabrication of complex designs that consumers wanted in their silk clothing, tablecloths, and wall hangings, which could be ruined by the careless mistake of a fatigued worker.

Looms inspired computers, which initially stored information with mechanical gears, paper tape, punch cards before such modern developments as the optical compact disc (CD). The information on a compact disc, for instance, can control a computer by telling it what sounds to play (if a music CD and the computer has an application that plays sound files) or what instructions to execute (if an application CD). This example shows two different kinds of information, which computer scientists term "data" and "program."

The peg or holes in silk looms represented data, as did the information ("feedback") from sensors on the stealth bomber. Program information for the looms was in the techniques of the loom operators and was shared verbally (from master to apprentice). Program information for the stealth bombers was in computer applications and was shared magnetically (from development computer to on-board computer). While programs may be complicated—computer programs may have millions of lines of code—they are based on algorithms, which are defined as the rules for solving a problem. For instance, a basic thermostat controlling a home heater, whose computer program would be completely unintelligible to most of us, follows this simple algorithm:

1. Measure current temperature (data)
2. Compare it to the desired level set on the dial (data from the human operator)
3. If current temperature is less than set temperature, then turn on furnace; otherwise, turn it off.
4. Loop back to step 1

Something common in bathrooms gives us another example. The Sonicare ™ electric toothbrush, which promises to clean and whiten your teeth with high-speed vibration of the bristles, follows this algorithm:

1. If the button is pressed, run for two minutes before turning off.
2. If the button is pressed while running, turn off.

3. If off for less than 45 seconds (you just wanted to add more toothpaste, for instance) when the button is pressed, run for whatever was left of the original two minutes.

4. If off for more than 45 seconds or placed back in the charger before the button is pressed, run for a full two minutes.

We find a slightly more complicated, but also more familiar, algorithm in the cookbook *Laurel's Kitchen: A Handbook for Vegetarian Cooking and Nutrition.* Their recipe for cornbread:

Ingredients	Instructions
2 cups cornmeal	Preheat oven to 425.
½ cup wheat germ	In a large bowl stir together dry ingredients.
1 teaspoon salt	In another bowl, mix the wet ingredients.
½ teaspoon baking soda	Combine the two just until they are well mixed.
1 teaspoon baking powder	Turn into an 8" x 8" baking pan, well greased.
1 tablespoon brown sugar	Bake for 20 to 25 minutes.
1 large egg, beaten	
1 tablespoon oil	
2 cups buttermilk	

Notice that the recipe does not specify who is following the rules. It could be a woman, a boy, a robot, or some brilliantly coordinated insects. The technical phrase for this flexibility is "substrate independence," but we can also view it as separating function from form, procedure from implementation, or information from matter. Because of the algorithm's substrate independence, understanding the behavior of a technology can often transcend the details of how that technology is implemented. Even when we comprehend little else about a technology, algorithms can allow us to predict its behavior.

Computers have been built on substrates of mechanical gears, magnetic relays, vacuum tubes, discrete transistors, integrated circuits of millions of transistors, test tubes of DNA, and even Tinker Toys™, the children's toy of wooden blocks, pulleys, strings, and sticks. Daniel Hillis, a computer scientist who designed one of the most advanced computers of the 20th century, created a Tinker Toy computer to play tic-tac-toe. It had neither the power nor the speed of the commercial products he designed, but it operated on the same algorithms. He summarized this substrate independence for computers: "One of the most remarkable things about computers is that their essential nature transcends technology." One of the most remarkable things about technology is that its essential nature transcends matter. Information in the form of algorithms is why.

We can apply this principle to understanding nanobots, microscopic robots built on the nanometer scale (each part of the robot might be just billionths of a meter or a few atoms in a row). Because nanobots have not been invented yet, we can only speculate about how they would work. Some call them science fiction; others predict they are a likely consequence of the nanotechnology we are already developing (which is still primitive, limited to such inventions as sunscreen with nanometer scale titanium dioxide particles to better block ultraviolet rays and Eddie Bauer pants that won't stain even with red wine). The consequences of nanobots are interesting enough and serious enough for us to start thinking about them just in case they are possible.

Free Matter and Valuable Information

Fabricating from the atoms up—perhaps even scavenging the carbon atoms with which we have polluted our atmosphere through smokestacks and vehicle tailpipes—with nanotechnology suggests that information may become more valuable than materials. To explain, we start with a couple technologies that already exist.

The value of a compact disc (CD) depends almost entirely on something invisible to the naked eye. Suppose you receive one in the mail. How much is it worth? If it is a new computer application, it could be hundreds of dollars. If a music CD, perhaps $10 to $15. If a pitch for Internet service (and you are already satisfied in that way), it is worthless—at best it would make a shiny drink coaster on your coffee table. If it contains the human genome, then it cost hundreds of millions to create, but since it is freely downloadable, the CD is not worth much. If you could only take that CD back a few years in time, you could sell it to those about to spend all that money on the human genome project! In short, the value of the CD you hold in your hand has much more to do with the information on it than in the materials that compose it.

New digitally controlled chemical molding systems may soon download information on the design of a table or bookshelf and then have the item pop out of an automated factory. Current designs of this "automated factory in a box" already produce hulls for boats, but could produce almost anything that can be molded out of plastic. Soon it may be possible to drop these from airplanes into remote areas suffering from natural or human disasters. They could produce sections of irrigation pipeline in the morning, download new plans, and produce containers to store food in the afternoon.

In the future, nanotechnology may remove the restriction that the product be molded out of plastic and make the automated factories small and affordable. A "nanotechnology factory" may use carbon as a raw material since it can be very hard (e.g. diamonds and fullerenes) and there's a lot of it just floating around in the air from all the fossil fuels burned the last few centuries. It would be possible to download the plans or specifications for an object and have it fabricated in your home or office by a "matter compiler." If that technology develops, our attitudes about information and material in products will be very different. And information may be one of the few things left with value.

Nanobots could manipulate matter at the nanometer scale, which means they could arrange and rearrange atoms. Scavenging atoms and molecules from their surroundings, they could make copies of themselves. The Sorcerer's Apprentice segment of the movie *Fantasia*, when Mickey Mouse lets replicating brooms get out of control suggests what could happen with nanobots. Once a nanobot makes a copy of itself, both it and the copy make copies. Then those four each make copies. Even if making a copy took a day, there would be one billion of them at the end of a month and 1153 quadrillion at the end of a second month. What's to stop this?

Perhaps an algorithm. The self-replicating nanobot could include a counter, which would start at "10" in the original nanobot. Then, when a copy is made, the counter decrements to "9" and the copy has its counter set to "9", too. The four nanobots in the next generation each have counter values of "8". When the counter reaches "0", the nanobots stop multiplying. This would result in 10 generations or 1024 nanobots. If this is not enough for our purpose, we could program the counter to start at a larger number. Even without understanding the substrate or the implementation of nanobots, we can understand this algorithm.

. This is not a new technique. Nature already came up with this technique to prevent rampant replication. Our algorithm is similar to that in some cells, which divide until their counter—a steadily shortening tail of *telomeres*—runs down. Now we get to apply something we know about in one field—medicine—to technology. This allows us to anticipate what might happen if the counter fails. Cancer is a disease that defeats the telomeres countdown, mutating the cell to allow unlimited reproduction. Could something like cancer afflict nanobots? Could a mutation in a self-replicating nanobot change its algorithm and allow it to replicate forever? What would stop it from converting all material on the surface of the earth into copies of itself? Evaluating nanotechnology while nanobots are still science fiction would seem a good idea.

Are Computers Really Getting Cheaper?

Near the end of World War II, the first general purpose electronic computer was built for almost half a million dollars. In 2004 dollars, that would be more than $5,000,000, and a 2004 computer costing $1000 would run circles around the Eniac. This is no surprise; computers have plummeted in price and continue to become more powerful every year (doubling in power every 18 to 24 months). What might be a surprise is that if we built just one new computer, rather than enjoying the economy of scale from making millions, it would cost more than the Eniac.

The Eniac is like an underdog coming into a boxing match. It's slow, able to execute just 5000 instructions per second. It's weak, with just 19,000 vacuum tubes and 1,500 relays. It's hungry, consuming almost 200 kilowatts of electrical power. And it's heavy, weighing over 30 tons.

A modern microprocessor would be the heavy "odds-on" favorite in a fight. It's fast, executing a billion instructions every second. It's powerful, orchestrating tens of millions of transistors (functionally like vacuum tubes or relays). It's not hungry, drawing less than 100 watts (double that for a complete PC). It's light, weighing less than an ounce (though, to be usable, it needs the rest of a 20 pound PC around it).

At first, the microprocessor wins on cost, too, at about $150. But suppose we made just one microprocessor. Research and development for its design would cost between $1/5$ and ¼ billion dollars. Plant, property, and equipment for a "development" fabrication plant would cost about half a billion dollars. That single microprocessor would cost about ¾ billion dollars! The Eniac seems almost a bargain at $5,000,000...as long as you don't need something more powerful than a handheld calculator.

Of course, nobody makes just one microprocessor. Instead, a company like Intel will invest almost two billion dollars more (than the ¾ billion we already counted) on plant, property, and equipment for a "production" fabrication plant and then produce 10 million processors each month. This makes possible massively powerful computers composed of many microprocessors. These include ASCI Q (12,000 processors), Blue Gene (one million processors), and the protein folding simulation project Folding@Home (more than 200,000 processors distributed all over the world).

Nanocubes are just a millionth of a millimeter
(a nanometre) across. Stacked like bricks,
they could make up a range of materials
with useful properties such as light emission
or electrical conduction. Many chemists are currently
trying to develop molecular-scale construction kits
in which the individual components are single molecules...
– Philip Ball

Organization Part 2: Repetition & Layers

Nanotechnology promises to be able to create nearly anything just by assembling atoms. That assembly would rely on two things: repetition and layers. Repetition is doing the same thing again and again, such as stacking brick upon brick. The layering that interests us is not the physical layering of bricks but the conceptual layering of bricks to make walls, walls to make buildings, and buildings to make towns.

The designer of a town need not understand how a building is made, the designer of a building need not understand how a wall is made, and the designer of a wall need not understand how a brick is made. Like Dr. Seuss' Cat In The Hat (with smaller cats under each hat), an onion, or a Russian Egg (with smaller eggs inside each egg), towns, buildings, walls and bricks are layers containing layers.

Biological processes layer with organs, cells, proteins, and molecules, using mass repetition at each layer. Technological processes do the same with their own building blocks. For instance, nanotechnology is the creation of objects at the nanometer (one billionth of a meter or roughly molecular) scale. If we could repeat and layer molecules, we could create just about anything for which we had a design. If we could do it economically, it would change our world. Although nanotechnology is very young, repetition and layers have already proven useful in a variety of complex technologies.

A billion microprocessor instructions execute in less than a second. In fewer than seven seconds, a microprocessor could execute one instruction for every human on earth—and that's on a machine costing just $1000. Who could possibly think up 7,000,000,000 different instructions? Nobody has to because most computers use about 100 instructions, which they repeat in many combinations. The power of the computer's language is in repetition.

A similar example of repetition can be found in robots. In the movie *Terminator 2*, Arnold Schwarzenegger faces a futuristic robot made of liquid metal. It can reshape itself to appear as a person or an object. Blow a hole in it and it heals. Smash it into pieces and the pieces flow back together, like mercury, to reform the original. Complete magic? Not if you imagine tiny robots—perhaps as small as nanobots—working together like a cheerleading team, but instead of forming a pyramid of a half-dozen people, millions of them would assemble into anything for which they had a plan. That plan or design would be encoded in each robot's software or downloaded from some other system.

The U.S. military is already testing robots just one millimeter on a side that are strewn from robot planes. They wait and detect passing vehicles, transmitting what they notice up to the plane, which relays it to battlefield commanders. One-millimeter robots are small enough to hide in carpets. How long before they can go beyond listening and signaling to choreograph their own movements?

One mobile robot, the PARC Polybot, transforms itself from a looped tractor tread for speed to a caterpillar for climbing or descending obstacles (like stairs) to a four-legged spider when the terrain is level but uneven. How does it do this? The robot is composed of a dozen identical modules that can attach to each other in many ways all on their own, using infrared communication between the modules and computers in each module. More and smaller modules make the robot more versatile. Plans for a 200 module robot will come closer to, but still well short of, the technology portrayed in *Terminator 2*.

Repetition in robots means greater tolerance to individual module failures and more flexibility in assuming shapes. Since the modules are all the same, they can be mass-produced and, so, lower in cost than similar modules that are unique.

Supercomputers made of many mass-produced microprocessors, cost much less than comparably powerful supercomputers based on older designs. The most powerful computers now consist of thousands of microprocessors in tight networks: ASCI Q will use 12,000 to simulate nuclear weapons at Los Alamos National Laboratory. The Blue Gene computer will simulate how proteins are created using more than a million processors.

Repetition is behind the power of printing with interchangeable type. In the 15th century, Johannes Gutenberg created a simple cast to form individual letters and a rack for holding them. Although the 11th century Chinese alchemist Pi Cheng developed movable type four centuries earlier, he had to deal with thousands of symbols. Writing in a language with just a few dozen symbols (without **j**, **v**, and **w**, the German alphabet then had just 23 letters), Gutenberg could better exploit the power of repetition. For this and a few other reasons, he was the first to make printing with movable type practical and successful.

Who would guess the power of repetition simply by looking at bins of metal blocks, each with a letter in relief? Yet, Gutenberg's press, distinctive because of these interchangeable symbols, enabled Martin Luther to give birth to the Protestant movement, scientific knowledge to spread, and the common person to read.

The power of repetition also enabled further specialization because it allowed writers to reach an audience far beyond their own town. They could research and write on subjects of interest to only a small percentage of the populace because a small percentage of Europe's population was still large enough to justify a book.

Specialists develop complex technologies by working on different aspects of it. In the opening of this chapter, we listed a few of the specialists necessary to understand a modern car, and this team approach is necessary for many of our technologies. How do we team up, break the complex into specialty pieces and then reassemble those into something that works? By *layering*, a technique that systematically hides information.

Imagine a modern computer as having layers like an onion, from the outer husk down to the core. We start with the application software (e.g. a web browser, email reader, accounting program, or word processor) seen on the computer screen:

1. **Application software** runs on...
2. **Operating systems**, which are programmed in...
3. **Machine languages**, which are interpreted by...
4. **Microprocessors**, which are assembled from...
5. **Logic gates**, which are comprised of...
6. **Transistors**, which are built from...
7. **Silicon** (with impurities).

Why layer? Because you can specialize in understanding one layer, using pieces of the layer below as if they were like Lego™ building blocks. You do not need to understand how to make them or how they work internally. All you need to know is what functions they perform. Someone working at the next layer up can use whatever you create, and that person need not understand the details of your design if you document its function.

This way of hiding information that's not needed is called "functional abstraction" because, instead of understanding all the details of the layer below, you satisfy yourself with a summary or abstract of what function or service it performs for you. The motivation for layering is making complex systems simpler to understand, design, build, test, and modify.

The concepts of repetition and layering offer good news in our quest to understand technology. If we decide we want involvement in some aspect of a complex technology, we may need to learn just one instance in one layer. Repetition allows us to replicate the one instance we understand with, often, predictable results. With layering, we need not swallow technology whole, but can bite off just a little by treating the layers above and below our bite as hidden. This means we need only understand how the layers neighboring ours behave externally. We do not need to understand the complexities of how they operate internally.

But incorporating these powerful concepts into our understanding of how technology works is like climbing a mountain from whose peak we can now see a higher peak. Well, two higher peaks.

The first peak has to do with the ability to communicate. If we are seeking expert knowledge in order to evaluate a complex technology, we may need to call on a different expert for each layer. An expert on one layer may have little to tell us about another layer. He may even use different terminology and, so, may have difficulty communicating with experts on other layers. It should not be a surprise, then, that layering falls short of eliminating all the challenges of understanding complex systems. Even great tools don't solve everything.

The second peak of our conceptual mountain is a bit of a surprise, however. The premise of our brilliant divide-and-conquer approach is that we can break complex technology down into its components and, by understanding one or a few of them, extrapolate to the whole. The key sentence three paragraphs back was, "Repetition allows us to replicate the one instance we understand with, *often*, predictable results," and the problem is in the word "often."

When the components being replicated are organized into certain types of networks, new behaviors emerge quite unlike the behavior of any of the components. This "emergent behavior" makes such systems hard to predict. While it is appearing more frequently in new technologies, it is already as common in Nature as ants.

> The amazing feats...comes not from
> complex actions of separate colony members
> but from the concerted actions of many nestmates
> working together...One ant alone is a disappointment;
> it is really no ant at all.
> – Bert Holldöbler

Emergent behavior

Peter Cochrane played with ants in a gutter, watching how they work and interact. Running back and forth on six tiny legs, ants have been perfecting their roles since the time of dinosaurs 100 million years ago. How many mistakes have they made? What is it about their behavior that is so effective that it has become dominant? Cochrane had good reason to be curious. As head of research and development for British Telecom, he had nearly 1000 people working to improve a vast network of computers, satellites, and cables. He believed that his people could learn a lot from ants.

Ants teach us how complex behavior emerges from a system of simple parts. An ant has about 200 neurons of brainpower, a bit less than the human complement of 100 billion neurons (Vision and hearing require lots of neurons to decode, so the 125 million neurons in the human eye may suggest why the 200-neuron ants rely more on smelling than seeing). If you were given an individual ant to study, it would be difficult to predict the behavior of a colony:

Hierarchy – The queen, larvae, workers (sometimes several castes and sizes), and soldier members identify themselves and each other with chemicals (pheromones).

Communication – Ants touch each other and use pheromones to convey about 10 to 20 equivalents of words or phrases. They communicate discoveries of a food supply, a new territory to explore, an invasion by enemies, or a good location for a new nest.

Career planning – Pheromones from the queen inhibit all but a few daughters from laying eggs, because the colony needs more workers than rival queens. Soldiers spread pheromones that, in sufficient quantity, cause larvae to develop into workers instead of soldiers, maintaining balance between the castes. After all, soldiers need to eat and it is workers that bring food.

Domestication – Some species of ants tend aphids much as we tend cattle because aphids secrete "honeydew" (actually excrement) rich in sugar, B vitamins and minerals. The ants protect the aphids fiercely and keep aphid eggs alongside their own. They know which parts (e.g. root or leaf) of which plants each type of aphid likes, carrying the aphids from inside the nest to the appropriate "pasture" or, when necessary, on to new pastures.

Slavery – While division of labor between workers and soldiers is common within a species, some ant species are equipped for nothing but attacking and kidnapping other species of ants. They rely entirely on their slaves for tending the eggs, gathering food, and repairing the nest. When not raiding another species' colony for eggs to raise as future slaves, they sit idly or beg food from their slaves. Deprived of the slaves, they would die.

Much of an ant colony's behavior emerges from the interaction of its parts. The simple rules that govern the individuals often lead to the appearance of coordination. Foraging ants leave a chemical (pheromone) trail. When they find food, they retrace their steps to return to the colony. In many colonies, ants follow the simple rule of following paths other ants have taken. As more ants find the food and return, the path gets a stronger chemical dose and becomes more attractive for other ants to follow. When the food is consumed or disappears, the ants go back to aimless wandering.

Observing from above, it might appear that the ants know how to find any nearby food and then coordinate legions to retrieve it. But without supervision or central coordination, what appears to be intelligent behavior actually emerges from nothing but the complex interaction of simple parts. This collective behavior is called *swarm intelligence.*

With simple rules and without supervision, ants can be very efficient. And inspiring: after building an ant farm in the British Telecom R&D laboratory, Cochrane had a team rewrite network software from 1,600,000 lines of code to just 1,000. Southwest Airlines copied ant colony behavior to gain more than $10 million in their freight delivery business. Transferring freight among their airplanes uses rules similar to those used by ants to forage. Finding the best path to food works along similar rules, it appears, to finding the best path to deliver a package. Swarm intelligence may be neither conventional nor intuitive, but it is effective.

Swarm intelligence is one example of what is generally called emergent behavior. If we build a system of parts that interact with each other, we may see it act in surprising ways, doing things that would be hard to predict from the behavior of each part. Is it the nature of all technology systems to exhibit this emergent behavior? Why don't we see it in toasters or in cars? What about technologies whose rules are far more complex than the rules foraging ants use? Will we have difficulty predicting how these "systems of parts" behave?

Not all complex systems exhibit emergent behavior. There are four ingredients to a system that does:

1. No central control
2. Parts work on their own
3. Parts affect each other
4. Effects can loop back to their source

Traditional technology is designed for clear and largely unchanging specifications. It is efficient but brittle. But, taking a cue from biology, technology is starting to adapt. As we attempt to solve problems that defy clean definition or problems that, by their nature, continue to evolve, we are designing more flexible technology. And we are finding that it tends to exhibit emergent behavior.

> We used to think that
> if we knew one, we knew two,
> because one and one are two.
> We are finding that we must learn
> a great deal more about "and."
> — Sir Arthur Eddington

Emergent behavior will appear in technology more frequently as we make it more complex and flexible. The Internet is a good example. It adapts and evolves to many environments around the world, having been designed by the U.S. government to adapt to the damage that could be inflicted by war. Attempts to block pornography, copying of music, or politically unacceptable information are all treated by the Internet as damage to the network. By its very design, it attempts to "heal" this damage. Its behavior may be difficult to predict, but it is flexible and robust.

Some Systems Work Even When Broken

Complex systems exhibiting emergent behavior are often tolerant of small failures. Stepping on a few ants does not cause the colony to fail. By contrast, removing a few lines of software could easily cause a computer to crash. Why are ant colonies so resilient? If the answer lies in their organization, can we copy it to make our technology resilient, too? The secrets behind resilient systems could be of value in many fields, as Albert Laszlo Barbarasi states in his book *Linked: the New Science of Networks*:

"Robustness is a major concern for biologists, who want to understand how a cell survives and functions under extreme conditions and frequent internal errors. It concerns social scientists and economists addressing the stability of human organizations in the face of famine, war, and changes in social and economic policy. It is a serious issue for ecologists and environmental scientists, motivating ambitious worldwide projects to preserve the sustainability of an ecosystem threatened by the disruptive effects of industrial development. Achieving robustness is the ultimate goal for specialists in increasingly interdependent communications systems, which must maintain a high degree of readiness despite inevitable malfunctions of their components."

Networks with many interconnections allowing feedback in a variety of ways, can form robust systems that can adapt. Coming from a different angle, this is beginning to describe the very same systems that exhibit emergent behavior, focusing on our 3rd (parts affect each other) and 4th (effects can loop back) ingredients.

Biology has evolved a pattern of organization that may become increasingly common in future technology, as we seek to replace our specialized, efficient, brittle systems with those that can adapt to damage and change. Some have called the 20th century that of the computer and the 21st that of biology. Computers are not going away in the 21st century, but they may start emulating the patterns we are discovering in biology.

Already the following views are widespread:
thinking is a type of computation, DNA is software,
evolution is an algorithmic process. If we keep going
we will quietly arrive at the notion that all materials
and all processes are actually forms of computation.
Our final destination is a view that the atoms
of the universe are fundamentally intangible bits.
— Kevin Kelly

Does technology work like Nature (e.g. an ant colony)? Or, visa versa, does Nature work like technology (e.g. the computer)? Kevin Kelly, founding editor of Wired Magazine, suggests this second approach, putting thinking, DNA, and evolution in computer terms.

The models we use to understand our surroundings affect what we see and how we make decisions. But most people do not even have models for how technology works. For them it is simply a mystery because it is as complex and detailed as the tinkering of engineers, technicians, hobbyists, and those who delight in disassembling technology to figure out what makes it tick. We need those people to keep things running, but most of us do not need that level of detail in order to understand how technology works.

Recognizing that we do not, allows us to evaluate the technology influencing many of our decisions. The truth is that simple and easily understood patterns are common to many technologies, and will probably apply to future inventions, as well. In this chapter we made a start, uncovering these patterns:

- *Energy* showed us that any technology, no matter how advanced, will rely on some form of energy, and may convert it into other forms. So, when trying to understand how technology works, we can look for a metaphorical power cord.

- Technology can be ***distributed*** or ***centralized***. Factors affecting which way it is organized include technological capability, cost, maintenance requirements, reliability of alternatives, and social concerns.

- *Control* systems monitor and adjust temperature, antilock brakes, airplane wing surfaces, and other "real world" systems. Feedback and correction are critical to many technologies.

- *Information* is the invisible component of technology. Software, crucial to every computer on earth, is no more than information. As material fabrication technologies advance, information in the form of algorithms and designs will become even more important.

- Complex technologies are often composed of ***repeating*** simple components many times and of concealing complexity within ***layers***. This means that we do not need to understand an entire system, but rather can focus on just one layer. It also means that evaluating a complex technology could require relying on a different expert for each layer.

- ***Emergent behavior*** warned us that understanding how the components of a system work does not necessarily tell us how the whole system works. As technological systems become more complex, we should be prepared for some surprises, and consider that in our evaluation.

Understanding how technology works illustrates a difference between competence and literacy. The engineers, technicians, and hobbyists possess, at least, competence. Those who grasp patterns spanning many technologies have an important component of literacy.

Chapter 5

How does Technology Change?

To fathom the changes
our marvelous tools will undergo,
we must draw on insights from
those who study evolution
of life and society.

If technology did not change, this book might not be very important. Or interesting. Early in human history, glaciers melted faster than technology changed—or at least both processes were comparably imperceptible to us. The first known plow dates back prior to 6500 years ago, after which more than 1000 years passed before the first known wheel…plenty of time for people to forget that there had once been a time before the plow. And then the wheel probably crept onto the scene slowly and inconspicuously.

Even if there had been newspapers, no headline would have read, "First Plow, Now Wheel: Change Rocks Civilization." No, this book and its search for patterns underlying the vast parade of technology would not have found an audience. Technology was then a static feature of the environment and patterns in its nature, change, and evaluation must have seemed irrelevant. But times change and technology accelerates, so we find ourselves facing…

- New weapons and new protections
- New poisons and new medicines
- Newly-obsolete industries and new careers
- New assaults on our privacy and new forms of creative expression.

We need not remember 1000 years back because less than a century ago the world had no SCUD missiles, early-warning radar, weaponized anthrax, gene therapy, computer industry, TV commercials, or jazz quartets that burn and sell their own CDs. The faster that technology changes, the more likely we are to ask questions about it.

How do we explain, for instance, how technology becomes so inconspicuous? Cellular telephones, once as magic as Dick Tracy's wrist radio, are now completely unremarkable…unless one rings in a quiet theater. Some technology seems to be a solution in search of a problem. Isn't necessity supposed to be the mother of invention? In cases where the problem is clear, what causes some solutions to be adopted and others not?

How do we explain the blistering pace of improvements in electronics? Today we can buy a computer that is twice as powerful as one costing the same just a couple years back. Is this an anomaly of our time? Have any other technologies followed this pattern? What can we learn from science and the study of biological change—evolution—that will help us understand technological change? In this chapter we answer these questions by exploring five patterns:

1. ***Disappearing technology***: Technology can become so common that we stop noticing it (e.g. light bulbs). It can also be built into other technologies so we can't see it (microprocessors).

2. ***Necessity's mother & daughter***: An old saying goes, "Necessity is the mother of invention," but the gasoline engine and the microprocessor show us that invention can come before we even figure out the need.

3. ***Advantage, prestige, compatibility***, and ***visibility:*** What influences whether a new technology is adopted or not? We find out why printing with moveable type succeeded with Gutenberg's press in the 15th century, but failed several millennia earlier.

4. ***Autocatalysis***: Technology acts on itself to change itself, accelerating even to an exponential degree. This applies not only to computers, but also mechanical clocks and biology.

5. ***Evolution & memes***: Natural selection guides both biological change, based on genes, and technological change, based on memes.

And while the stories are interesting, the patterns will help us understand and evaluate today's technology, and—when it comes—tomorrow's, too. That last part is an important goal of this book: learning enduring patterns so that we can more consciously create our future. This is different from predicting the future—something far more difficult.

For instance, Microsoft failed to predict the importance of the Internet, forcing its legion of brilliant people to play catch-up with Netscape's web browser. In 1948 Thomas Watson, then head of IBM, predicted "a world market for about five computers." In 1977 computer pioneer Ken Olsen claimed, "There is no need for any individual to have a computer in his home." Predicting specific technologies and applications is a gamble. Influencing our future need not be.

We can make informed, thoughtful, critical, evaluative decisions throughout our lives—and, collectively, that is how we create our future. Context is the foundation for those decisions. Whether something new matches or differs from one of these patterns, the context we gain by recognizing those patterns makes it easier to identify, categorize, and understand the new. Let's start with something familiar, something that you or I would not think twice about, but that stunned the Emperor of Brazil.

Disappearing Technology

In 1876, Dom Pedro, Emperor of Brazil, presided over the Centennial Exposition in Philadelphia. His long trip was made worthwhile when he saw Alexander Graham Bell's demonstration of the first telephone. He was stunned, but so was everyone else. Squeezing someone's voice into tiny metal wires, only to have it pop out on the other side of the building, was magic. But this was not a trick; it was the foundation for an industry.

A few years later, on New Years Day, 1880, crowds traveled to Edison's laboratory in Menlo Park, New Jersey, just to see his new incandescent bulbs light up. Today, telephones and light bulbs blend into our landscape. Who even notices them?

In the 19th century an airplane passing overhead (years before the first one was invented) would have caused excitement and panic. Today one may notice the contrail from a jet, but even wide-eyed children soon learn to ignore the regular passing of passenger, cargo, and military aircraft.

Three centuries ago, England's Parliament offered a prize equivalent to several million of today's dollars for the invention of a timepiece that would work on a sailing ship. Both highly accurate time and star charts were necessary for locating oneself at sea, as missing your destination, ending up lost at sea or on a strange and uninhabited shoreline, could mean financial disaster or even death.

John Harrison, a clockmaker, invented the first clock impervious to the rocking and rolling of a ship. This was a tremendous accomplishment in the 18th century, but today, how many of us notice that we have a descendant of these miraculous devices strapped to our wrists?

How Quickly Technology Becomes Popular

Familiarity breeds a form of blindness, and technology is becoming familiar ever more quickly. The table below shows how many years before a new technology was adopted by 25% of the U.S. population. Late 19th century technologies such as electricity (which we date from when it was first available for purchase), the telephone, and the automobile took decades to reach the masses: 46, 35, and 55 years, respectively. But late 20th century technology such as the personal computer, cellular phone, and the World Wide Web took about a decade: 16, 13, and 7 years, respectively. When every fourth person owns something, it is not surprising that it tends to blend into the background.

Product	Year Invented	Years to Spread
Electricity	1873	46
Telephone	1876	35
Automobile	1886	55
Airplane	1903	64
Radio	1906	22
Television	1926	26
VCR	1952	34
Microwave oven	1953	30
Personal computer	1975	16
Cellular phone	1983	13
World Wide Web	1991	7

Take a look around you. How far back would you have to travel in a time machine with some of these familiar objects in order to be called a magician or witch? How long will it take for today's most advanced technology to become commonplace? Of course this pattern of "familiarity leading to a form of blindness" is not specific to technology. The new catches our attention because, as animals, we are wired to notice rapid changes in our surroundings (e.g. hungry leopards). The new could be a threat or an opportunity, and it takes attention to evaluate it.

However, there is something different about how technology disappears. A bizarre new hairstyle or a practice like body piercing may become so common that it blends into the background, but technology is actually burrowing out of sight. Physical connections between telephones (i.e. cords) are disappearing with the increasing use of cellular phones and wireless computers. Microprocessors are disappearing into the engines of our cars (adjusting fuel and air mixtures, operating the antilock brakes, etc.) and the doorknobs of our hotels (for which you use a magnetic or punched card instead of a conventional key). Why is this?

One reason is that technology is assembled from parts and those parts are often other technologies. The microprocessor becoming a building block for cars fits this pattern. But there is more to this "burrowing out of sight" than that. In the chapter on why we use technology, we saw that our reasons for using it last much longer than the specific technologies themselves. Beyond the novelty value—which includes the temporary status that flaunting some new device may confer—most of us just don't care about the technology. It might as well disappear and get out of our way.

Future bathrooms may recognize who enters, check their vital signs (e.g. blood pressure), remind them about taking medicine, and even contact health care providers. It may also monitor health by analyzing what we leave in the toilet. Accenture (a technology and services consulting firm) developed a prototype bathroom of the future that incorporated many of these features.

With its *ubiquitous computing* strategic initiative, IBM seeks to make computer technology hide in anything from milk cartons to bottles of prescription drugs, from sprinklers to door locks, from clothing to the toilet. A company called emWare produces $1 computers just intelligent enough to communicate over the Internet.

Connected computers like these, hidden in everyday objects, are now joining billions of *embedded microcontrollers* that have long established themselves inside factory machines, elevators, lighting/heating/cooling systems, and other places, and we have already stopped noticing them.

If future bathrooms analyze our toilet deposits to predict health problems, it will not be entirely new. Long ago, the Chinese Emperor had his stools checked by royal doctors to make sure he was in good health, but in our time there has been no visible groundswell for this luxury. If consumers are not demanding it, then why is someone inventing it? Isn't *necessity* the mother of invention? Interestingly, it is sometimes the daughter.

It was by chance that the recorded sound
found a lasting commercial use.
With the business market for the phonograph
faltering [wax cylinder recordings did *not* replace
business letters], manufacturers scrambled
to come up with other applications. In 1889,
the first coin-operated phonograph was placed
in an arcade. For a nickel, listeners could hear
a two-minute recording.
— Mark Robinson

Necessity's Mother & Daughter

The incandescent light bulb was invented before Thomas Edison was born. But it did not work well, which is why Edison, using improvements in vacuum technology and a few other tricks to make it practical, is credited with the first bulb. Widespread demand for safe and economical lighting made the light bulb an example of "necessity being the mother of invention," but history shows that the relationship is often reversed.

Edison invented the phonograph to record people's dying words or to replace business letters (which the earlier invention of the typewriter had started to transform). There was little demand for deathbed speeches or wax cylinder business correspondence (spoken letters). Instead, Edison lived to see entertainment, which he declared as frivolous, take over his invention.

Ordinary people dropping coins into phonograph machines to hear a brief song was hardly what Edison had in mind, yet that led to huge industries based on records, tapes, and, eventually, compact discs. Listening to recorded audio (as well as the very similar technology of recorded video) has become so popular and pervasive that it has become nearly a necessity…and the phonograph was mother to that.

The modern computer industry has its own example of invention being the mother of necessity. The microprocessor was invented in 1971 to power a calculator. When the price of calculators plummeted from $250 in 1972 to just $10 a year later, the microprocessor was an invention in search of a need. This orphan has since found its way into hundreds of millions of homes, hiding in microwave ovens, TVs, video games, thermostats, and stereos, as well as the obvious personal computers. Prior to the microprocessor, these technologies were doing quite well without it (e.g. thermostats that turned on when it got cold, not an hour before our alarm clocks ring) or simply did not exist (e.g. personal computers). Since then, consumers have come to appreciate the new features that microprocessors allow, and manufacturers have come to appreciate the cost savings of controlling their product with a programmable device.

A story with twists and turns between necessity and invention concerns the steam engine, driver of the Industrial Revolution. Steam engines require lubrication or else they overheat and seize. Necessity. Mineral oil was distilled from petroleum to be that lubricant. Invention. Here, necessity was the mother of invention. However, mineral oil also leads indirectly to the demise of the steam engine in a case of invention being the mother of necessity.

For years when distilling mineral oil from petroleum left a waste product: oil too light to be a good lubricant. This dangerously flammable liquid was dumped into rivers or anywhere convenient. In the 19th century, inventors were hard at work improving both steam and natural gas engines, so eventually someone thought of using this light oil—let's call it gasoline—for fueling an engine. Shooting drops of gasoline into an engine cylinder exposed only the surface of each drop to oxygen, something crucial to burning. Once the surface burns off, oxygen can reach a layer deeper, but that means a slow explosion going layer by layer, like peeling an onion. As it turns out, the solution to this problem was invented to solve a different problem.

Scientific experimentation with liquids led to the "spray atomizer," which was quickly adapted to create a fine mist of perfume. Bathing was not common before central plumbing and hot water heaters, making perfume understandably popular. The medical belief that disease came from bad smells made development of the atomizer urgent. But atomizing a liquid into many very small drops is also exactly what gasoline needs to get maximum exposure to oxygen, and therefore the quickest and most energetic explosion.

Once the atomizer was recognized as an important part of the solution, it was adapted and refined into a new invention, the carburetor (a form of which we still have in our cars). As gasoline engine technology improved, the steam engine lost popularity (there were other factors contributing to its demise, but we need to return to our point). Necessity was the mother of the carburetor; by the time it came about there was a clear need for improved efficiency in the very-promising new engine. Still, the two *grandmothers* of the carburetor were not necessities, but, rather, other inventions: a distilling process that just happened to produce an explosive waste product and a type of engine that could use it.

Which came first, necessity or invention? For technology in general, the answer is the same as for the riddle, "Which came first, the chicken or the egg?" The chicken came before the egg, which came before the chicken. Technology satisfies needs that develop in response to capability, which develop in response to earlier needs. Our Cro Magnon ancestors of tens of thousands of years ago had no need for telephones, resealable sandwich bags, or accounting software. They had their own needs of communication, food, and organization, for which they invented solutions, which revealed new needs, and so on all the way to the present day.

By recognizing the natural swing between necessity and invention, we are better equipped to understand current and future changes. The next time we see a frivolous, but remarkable, technology, we may start looking around for sprouting necessities.

The four stages of response to
any new and revolutionary development:
(1) It's crazy!
(2) It may be possible—so what?
(3) I said it was a good idea all along.
(4) I thought of it first.
– Arthur C. Clarke

Advantage, Compatibility, Risk, Visibility

If an invention can create its own necessity, why do some inventions become no more than an idea or prototype? One reason is lack of infrastructure. Leonardo da Vinci envisioned the helicopter, but the strong, lightweight materials and power source were centuries away. The helicopter reached no farther than his notebook. In the chapter on how technology works we saw that the stealth bomber's shape and dimension appeared decades earlier as the flying wing, but that without small, fast computers, the design was unstable in flight. The prototypes were not adopted by an air force. Is lack of supporting technology the only reason that inventions fail? Once an invention clears this rather obvious technical hurdle, it faces the social hurdles.

Advantage, compatibility, risk, and visibility—technical and social factors—determine whether a viable technology spreads or withers. The success of the transistor over the vacuum tube illustrates these four factors:

1. Transistors have the ***advantage*** of being smaller, requiring less energy, producing less heat, and lasting longer than vacuum tubes. A transistor is a solid sandwich of silicon or other semiconductor. A vacuum tube resembles its parent, the incandescent light bulb.

2. Transistors amplify or switch on and off electrical signals, which is just what vacuum tubes do, which makes them sound ***compatible*** with that existing technology. Actually, they were not because manufacturing vacuum tubes and designing them into radios and televisions were both very different for transistors.

3. That the transistor performs the same function as a vacuum tube does mean that it is simple to imagine replacing one with the other. This lack of complexity means that this new technology has a low ***risk*** of failure.

4. The transistor's advantages over the vacuum tube were easily ***visible***. The pocket-size transistor radio introduced by Sony in 1952 ran on batteries and became very popular. The much larger vacuum tube radio used so much energy that it had to be plugged into a wall outlet.

Transistors succeeded as a technology because they had advantages over vacuum tubes, those advantages were easily visible, and the risk of failure was low. These factors overwhelmed the lack of compatibility that transistors had with the vested interests of manufacturers and designers. So a technology need not triumph in all four categories in order to succeed.

In the case of the transistor, further advances showed how overwhelming its advantages truly were. Integrated circuits, invented a decade after the transistor, miniaturized entire electronic circuits of multiple transistors onto a single chip of semiconductor. Tens of millions fit into postage stamp sized areas and run for years without a failure.

By contrast, vacuum tubes cannot be miniaturized, in part because they use lots of energy, which produces lots of heat, which requires space for cooling. Their reliability has improved since the middle of the 20th century: on average, one out of the 19,000 vacuum tubes in the Eniac computer had to be replaced every seven minutes. But it has not improved enough to make practical a computer with tens of millions of tubes. Vacuum tubes do live on in giant form as cathode ray tubes (CRTs) in televisions and computer monitors. Ranging from 19 to more than 30 inches across, these displays were not threatened half a century ago by transistors, but are now by liquid crystal displays (LCDs), which have been used in wristwatches for decades.

These four factors are also a useful tool in understanding why the invention of printing with moveable type, thousands of years before Pi Cheng (11th century China) and Johannes Gutenberg (15th century Germany), failed. Unearthed on the island of Crete and dating from 1700 BC, the Phaistos Disk is a clay disk imprinted repeatedly with 45 different symbols. Each symbol appears identically every time, indicating that it was probably carved into wood or cast into metal. Then, each time that symbol was needed, the carving or casting was pressed onto the disk.

Why didn't printing with moveable type "take" way back then instead of waiting more than two millennia to encroach on writing by hand? Because the Phaistos Disk failed all four tests:

1. It lacked advantage because, with only a handful of people (scribes) who could read in each kingdom, there was little writing of any kind to automate and no market for mass-produced documents.

2. Controlling writing was in the interest of the scribes. Why would they want to make it something that just anyone could do by pressing a symbol into moist clay? This made printing incompatible with vested interests.

3. Carving or casting 45 different symbols into various materials to test the endurance of each was time consuming. Testing various types of clay and processes for firing it would have been complex. This increased the risk of failure and likely discouraged many people.

4. Advantages were not visible because the process of carving or casting symbols and then pressing them into the moist clay was slower than writing on it with a stylus. Visibility would ultimately come with the demand for and capability of high volume printing.

Many things changed by the 15th century. One of the most important was the availability of paper. Invented in China, the process of making paper traveled to the west, where it replaced vellum, the skin of lambs. A copy of the Gutenberg Bible would have consumed 150 lambs. Still, this "vellum" had been an improvement over hauling around heavy and fragile clay tablets. In Mel Brooks' irreverent movie *History of the World*, Moses comes down from the mountain with 11 stone tablets. Just as he is about to announce the "11 Commandments," one tablet falls and shatters. If only rag paper had been available! Paper was one of many reasons that Gutenberg's printing press succeeded:

1. *Advantage*: With thousands of literate monks spending their lives hand-copying manuscripts and many merchants in need of documents to facilitate long-distance trading, mass-production of written documents had great economic advantage. Just the century before, the Black Death (bubonic plague) had killed off cheap labor. Further, Europe was fascinated with machinery: clockworks, locks, and water gardens with fountains shooting up to do "tricks." A machine to automate printing had prestige.

2. *Compatibility*: The printing press was compatible with the values of the Church (printing Bibles, prayer books, and papal indulgences), the Protestant movement (spreading word of an alternative to the Church), and merchants (selling books of all sorts).

3. *Risk*: Developing an infrastructure for mass printing would be time-consuming, complex, and, therefore, risky if done from scratch. But, by the 15th century, papermaking had spread from China, olive presses could press inked letters into paper, and metal-

lurgy developed for clocks and door locks could also cast perfectly standard letters and symbols. Risk was low.

4. ***Visibility***: People could easily observe that printing copies of the Bible was far better than hand copying (each of which could have taken a monk his entire life). Further, the fixed costs of printing could be quickly recovered. Setting up to print a book cost roughly three times as much as having a scribe copy it, so saving would be realized once just a few copies were printed.

But we are getting ahead of the story. Why did printing with movable type not flourish in 11th century China? Lack of advantage and compatibility. The Chinese alphabet has thousands of symbols. Casting them into stamps, storing several of each type (since a symbol can appear multiple times on each page), and setting up the press would have been an immense effort. An advantage of movable type is in replicating fewer symbols more times (the power of replication we noted in the previous chapter). Also, printing with movable type was not compatible with Confucian values, which prohibited commercialization of printing. The government in China could distribute documents, but merchants could not sell them.

Modern printing is done with computers, and we can look to a computer keyboard to see most of the symbols available. Where did the keyboard's layout come from? Why are letters, numbers, and other symbols placed where they are? The adoption of the present day computer keyboard illustrates a battle in which ***advantage*** and ***visibility*** lost out to ***compatibility*** and ***risk***.

The keyboard on nearly every computer is similar to the one that was on nearly every typewriter. Called the QWERTY layout because of the letters across the top left, it was intentionally inefficient. It originated because early mechanical typewriters would jam if one typed too quickly. Inefficient placement of letters slowed typists so that they would not jam the machines, which used many metal arms, each with a symbol (or two, when used in conjunction with the shift key) in relief, arranged in a semicircle. Pressing a key caused an arm to swing up and hammer its imprinted letter against an ink ribbon and the paper.

But by the time the technology was no longer subject to jamming (with electric typewriters and the no-moving-parts word-processing computers we use now), many people had learned the QWERTY layout. So changing to a more efficient layout, such as Dvorak, would require learning a whole new layout. QWERTY and Dvorak split our four factors:

1. ***Advantage***: Although QWERTY minimized key jamming on manual typewriters, Dvorak allowed much faster typing on electric typewriters and computers. *Dvorak wins.*

2. ***Compatibility***: Few typists could touch-type on Dvorak keyboards, but a huge labor pool with QWERTY skill was ready for hiring by the same companies that had to choose which keyboards to buy. *QWERTY wins.*

3. ***Risk***: Trying a new system such as Dvorak is risky because it requires changing lots of equipment and retraining lots of people. It is much safer to just stick with what has long worked, however inefficient it might be. *QWERTY wins.*

4. ***Visibility***: The U.S. Navy reported that Dvorak was so much more efficient that re-training would pay for itself within 10 days, an easily visible advantage…if a business goes to the risk of trying it for itself. *Dvorak wins*.

Just because each keyboard layout "won" two factors does not make the battle a tie. In this case, compatibility was so important that the technically superior Dvorak keyboard was marooned as little more than a prototype. Few use it and 21st century computer keyboard manufacturers continue to use the 19th century keyboard layout.

Dvorak Keyboard Layout

In cases such as keyboard layout, compatibility outweighs all other factors. Economists call this "lock-in" or "network effect," which is what happens when a technology establishes itself so strongly that consumers are reluctant to switch, even to an apparently superior technology.

Lock-in is one factor in the success of the Microsoft Windows™ operating systems and of personal computers in general: there are so many of them in the market that any new product—hardware or software—is bound to support them. Buying a niche operating system or machine exposes you to the risk that future add-on products will not be made available for you…and that the market might abandon your system entirely. Economies of scale also favor the entrenched party with falling prices, but even being entrenched is no guarantee when the economic advantages of a new technology are overwhelming.

Eventually, a technology with an advantage can threaten one with compatibility, as the inexpensive Linux operating system has come to slowly take some market share away from the omnipresent Windows. While it is impossible to predict the outcome of Linux vs. Windows, we do know how the 19th century "war of the currents" turned out.

Edison built the first electrical power plants to produce direct current (DC). This means that electricity always flows in just one direction, electrons coming out of one slot on your wall plug, going through your appliance to drive it, and back in the other slot. Nikola Tesla, an odd genius who went from rags to riches and back again, advocated alternating current (AC), which reverses the direction of electrons many times each second (60 times in the U.S.). The war between DC and AC had high stakes. Electricity would clearly revolutionize the world; selling it and all the devices that used it would be very profitable.

With Edison established as virtually a "god of technology" in America, he invented the many technologies necessary to efficiently generate electricity, distribute it, measure how much a customer used, and charge for it. Making matters worse for AC, electric motors that ran in factories needed DC. Everything was compatible with DC and very little with AC. What could change the status quo?

To answer that question, we need to explain a bit about electrical voltage, current, and energy. Voltage measures the force moving the electrons and current measures how many electrons are moving. The speed of a river is analogous to voltage and the quantity of water to current. The amount of electrical energy is the voltage multiplied by the current, so we can chose to have high voltage and low current or low voltage and high current without affecting the amount of energy. The practical application of this theory comes into play when sending electricity over a long distance. It turns out that the amount of energy lost to heat is proportional to the amount of current, so the farther a utility sends their energy, the higher voltage and lower current they prefer. On the other hand, high voltage can be dangerous since it can arc right through the insulation on wires, causing fires or painful shocks, so consumers want low voltage and high current.

Whatever the proportion of voltage and current, the amount of energy is the same, requiring the same amount of coal, oil, uranium, or flowing water to generate. But here's the hitch: transforming one voltage to another is much easier with AC than DC. Because he needed to transmit his DC at the same voltage as consumers used it, Edison located his generators within a few miles of them. AC provided the best of both worlds, allowing transmission from faraway sources at efficient high voltage (hundreds of thousands of volts) and household or business delivery at safe low voltage (120 and 220 volts in the U.S.). As AC technology developed, this presented a critical advantage that DC could not match.

Niagara Falls, the first big hydroelectric generator, transmitted high voltage AC to cities many miles away, making visible the advantage of AC. DC still thrives in short distance applications, such as inside electronics, which run on the DC from batteries or convert AC "from the wall" into DC. Just as QWERTY compatibility overwhelmed Dvorak advantage, AC advantage overwhelmed DC compatibility.

> Technology is autocatalytic
> in that it acts on itself
> to change itself.

Autocatalysis

When something acts on itself to change itself, as some technologies appear to, it is called autocatalytic. There is a web of interrelationships and dependencies that suggest technology works this way. Tools create better tools, which create still better tools. This happened with primitive stone, bone, and wood tools being used to make improved versions. The light bulb needed a good vacuum pump in order to avoid burning out. Integrated circuits needed photo resistive chemicals to etch complex patterns on silicon (those chemicals were discovered just a few years before the integrated circuit was invented).

Try going back in time to invent something and you'll probably be stumped by absence of support technologies. We saw some of this in the dead-end Phaistos Disk, the early attempt at printing with moveable type on clay tablets, but it is easy to imagine more difficult situations.

If Mark Twain's *Connecticut Yankee in King Arthur's Court* had tried to invent the electronic computer, he would have had to invent much of the electrical and electronics industries. So, too, would Daniel Defoe's *Robinson Crusoe*, stranded on a desert island without even the metallurgy and labor that the Connecticut Yankee might have found.

Although interrelationships of technology may prevent an invention from coming before its time, they also propel a variety of technologies faster and faster. Viewed mathematically, autocatalysis can cause exponential change. Exponential change became familiar to many of us toward the end of the 20th century in the form of personal computers doubling in power every few years, but costing the same or less. This repeated doubling can be traced back to the beginning of the 20th century, when computers were built from electromagnetic relays and vacuum tubes.

The graph below shows computing power per constant (1998) dollar from 1900 to 2001. In 1908, one calculation every 50 seconds cost $154,000 with Hollerith Tabulator, the computer used for the U.S. Census. In 2001, one billion calculations every second cost about one thousand dollars in a common PC. In 2004, manufacturers shifted their focus from ever-faster clock speeds to squeezing multiple processors into one package, continuing the trend.

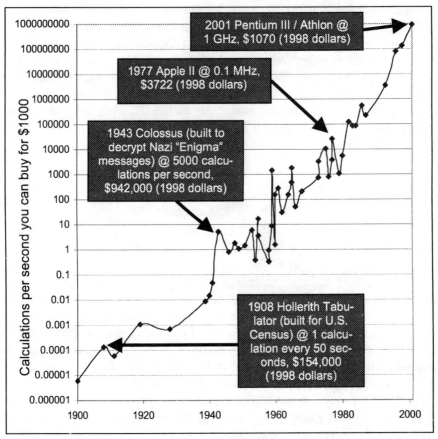

Suppose that growth from the Hollerith Tabulator and onward progressed linearly instead of exponentially. In 1909, a computer costing the same $154,000 would have done two calculations in 50 seconds. In 1910, three, and in 1911, four. By 2005, it would perform 98 calculations in 50 seconds, or less than two each second for $154,000. What a difference exponential growth makes!

Digital storage technology—computer memory and disks—has increased exponentially. Communications technology has, too, dropping the cost to send a given amount of information. Robotics and improved algorithms sped the sequencing of the human genome—early predictions had it taking thousands of years. We find similar exponential improvement in mechanical clocks, which were preceded by water clocks. One type of water clock, called a "clepsydra," consisted of a large bowl filled with water and a smaller bowl with a tiny hole in the bottom. The smaller bowl floated on the water in the large bowl until enough water leaked up to sink it. The final rush of water into the sinking bowl made a "ploonk" sound, indicating that, whatever you were doing, your time was up.

Water clocks had the unfortunate characteristic of clogging with sediment from the never-quite-pure water, which affected their accuracy. The first mechanical clocks—certified to be sediment free—lost about 15 minutes each day, the same as the best water clocks. But then they followed a curious trend. Every 30 years, or so, mechanical clocks doubled in accuracy, right up to the 20th century. Fashion in watch technology followed a similar trend with the thickness of British and Swiss pocket watches shrinking:

- 41 millimeters in 1700
- 25 millimeters in 1812
- 12 millimeters circa 1815-1825
- 11 millimeters in 1846
- 6 millimeters in 1850

But exponential change in clock accuracy and watch thickness did not continue, nor did faster air speeds. Over nearly a century, the maximum speed achieved in air doubled every nine years. When commercial supersonic transport (SST) went into service in the 1970s, many imagined SSTs to be the future of air travel. Three decades of technological advance later, the successful airplanes are more economical, but not as fast. In April 2003, Britain and France announced that the only remaining SST, the Concorde, would no longer fly. So much for exponentially increasing speeds of air transport.

Given the time that all air passengers must spend just getting to and through airports, the shortened flying time was rarely worth the price. Unlike color television and the videocassette recorder, which also debuted as expensive toys for the rich, the cost of flying an SST did not plummet.

While some technologies are not autocatalytic, those that are—computation, storage, communication—will change the world. And even those that change more slowly are subject to selective pressures from their environment, and that may help explain why some thrive while others become extinct.

> Technology picks right up
> with the exponentially quickening
> pace of evolution.
> — Ray Kurzweil

Evolution & Memes

It took billions of years to evolve from the first single-cell life form to multicellular plants and animals, but only half a billion to get from there to mammals. Then, less than a sixth of a billion from there to monkeys and apes, and $^1/_{36}$ of a billion from there to simple technology. The calendar opening "Chapter 3: Where does Technology Come from?" showed, on the timescale of millions of years, just how recently important technology has been invented. For example, electricity in any form more practical than novelty is barely two centuries old. That is just one five-millionth of a billion years. It is as if an accelerating arms race among predators and prey spilled over from biology to technology:

Steps in evolution	When it occurred (millions of years ago)
First life on Earth	3,000
Multicellular plants & animals	700
Mammals	200
Monkeys and apes	30
Humans, simple technology (stone tools)	2
Agricultural technology (plow)	0.01
Communications technology (printing press)	0.0005
Computing technology	0.00006
Computing/communications synthesis	0.000009

We turn to evolution because that may influence both biology and technology. If we want to understand what is accelerating the change of technology, we need to examine how evolution works, and in particular natural selection or "survival of the fittest." While this has been associated with biology and genes, it is more universal than that, applying quite well to technology and memes.

Could it be that the same forces of evolution that have governed biological change on Earth also apply to technological change? It might be easy to jump to the conclusion that, lacking genes, technology cannot evolve. Our Stone Age ancestors might have concluded that, lacking stone, one could not have a knife. More recently, we might have concluded that a knife must have a sharp edge…until we saw how effectively a laser can cut through eye corneas or steel.

So let's step back from the biological implementation of evolution to understand its process. What is necessary for evolution by natural selection? The answer is just three things, none of which require genes: variation, selection, and retention. Applying these to biology, which does use genes:

1. **Variation** – Sexual combination, random mutations, symbiogenesis, and viral penetration of the nucleus can cause variation in genes. A simple, if unrealistic, example: a giraffe with a medium-length neck has short-, medium-, and long-neck offspring.

2. **Selection** – Survival of the fittest selects some organisms to procreate and pass on their genes before dying and others not. The short-neck giraffe dies before procreating.

3. **Retention** – Offspring retain part or all of the genetic code of those successful enough to procreate. The medium-neck giraffe tends to have medium-neck offspring and the long-neck giraffe tends to have long-neck offspring.

Symbiogenesis is the emerging theory that *variation* is largely the product of combinations of life forms. An example is lichen (as seen affixed to trees and stone), which is the symbiotic relationship of fungi and bacteria. The fungi store water and provide protection. The bacteria photosynthesize nutrients. If the branches of the evolutionary tree rejoin to create new species then, according to scientist Lynn Margulis, "Animal evolution resembles the evolution of machines, where typewriters and televisionlike screens integrate to form laptops, and internal combustion engines and carriages merge to form automobiles. The principle stays the same: Well-honed parts integrate into startling new wholes..." So we apply variation, selection, and retention to the steam engine, which does not use genes:

1. **Variation** – James Watt created a variation on Thomas Newcomen's primitive steam pump.

2. **Selection** – Industry selected Watt's steam engine over competing technologies, such as waterwheels.

3. **Retention** – New steam engines retained the basic design while introducing variations, such as Richard Trevithick's more efficient high-pressure approach, bringing us full circle.

The difference: biology stores the information necessary for evolution in genes and technology stores it in human brains. Scientist Richard Dawkins coined the term "memes" for a "unit of cultural transmission." Memes could include ideas, designs, practices, or even musical melodies. They spread by word of mouth, through books, in classrooms, and on television. They spread imperfectly, with some individuals perceiving a *variation* of the original meme. Survival of the fittest *selects* which memes will be repeated to friends and which will sell on the media. The meme we get *retains* part or all of the information from the original meme.

The specific implementation of an evolutionary system—whether genes or memes or something else—does not matter, so long as the system incorporates variation, selection, and retention. Another way to put it: this system is *substrate independent* (a term that has come up a few times in this book).

Variation

Selection

Retention

Variation, selection, and retention in imaginary giraffes

> Imagine a world full of hosts for memes (e.g., brains)
> and far more memes than can possibly find homes.
> Now, ask which memes are more likely to find
> a safe home and get passed on again?
> – Susan Blackmore

This is not a radically different way to view technological change. It is conventional wisdom that (1) people innovate, (2) some innovations thrive while others are abandoned, and (3) others copy the innovations that work. What is different is that we are suggesting that some of the analytic tools developed for biological evolution may be turned on technology.

Suddenly it becomes reasonable to ask if there are technological counterparts to weeds, parasites, and viruses. What about inbreeding and the weakness that results from absence of diversity? For instance, is the Internet more vulnerable to computer worms and viruses because 90% of personal computers run the Windows operating system? Is this analogous to the danger that bananas, now virtually all the same genetically, could be wiped out worldwide by a single pest?

Naturalists have learned much about ecosystems, and now that may shed light on the increasingly complex and diverse world of technology. But we have new tools to explore the world of ecosystems, tools that let us simulate environments and squeeze years into minutes.

Computer scientists have created primitive virtual environments inside computers, which introduce variation, select the fittest, and retain many of their characteristics. These simulated worlds of ones and zeros instead of molecules and energy are still very simple, but they make it possible to see what emerges after thousands of generations subject to the rules of natural selection.

> Two hundred years after [Benjamin] Franklin, artificially generated lightning—tamed, measured, and piped through wires into buildings and tools—is the primary organizing force in our society, particularly our digital society. Two hundred years from now, artificial adaptation—tamed, measured, and piped into every type of mechanical apparatus we have—will become the central organizing force in our society.
> – Kevin Kelly

A breeding experiment by scientist Danny Hillis provides an illustration. Now Danny is not a virologist, breeding bugs in test tubes, but a computer scientist breeding programs in a computer. He let his evolution simulation run for 10,000 generations. His digital "creatures" competed for resources and the most successful mated, passing their genetic code on to digital children.

Occasionally, that code mutated, usually creating monsters that could not survive. Once in a while, though, the mutation was beneficial and the mutant thrived, producing long lines of children. Simulating 10,000 generations is quick on a computer with 64,000 interconnected processors running simultaneously, yet it still evolved a program so "fit" that it was only one step longer than the leanest, most efficient one that computer scientists had spent decades devising and improving.

That program sorted numbers into order, but as simulated environments become richer, with further programming and increasingly fast computers to run them, all sorts of technology may be evolved rather than designed. Why would we do this? Hillis answers, "There are only two ways we know of to make extremely complicated things. One is by engineering, and the other is evolution. And, of the two, evolution will make the more complex."

Evolution is an extraordinarily powerful force that created dinosaurs and humans out of energy and stardust. What will we create in the miniature universes boxed under our desks? How will we decide what to release into the outside world?

Two billion years ago our ancestors were microbes;
a half-billion years ago, fish;
a hundred million years ago, something like mice;
ten million years ago, arboreal apes;
and a million years ago, proto-humans
puzzling out the taming of fire.
Our evolutionary lineage is
marked by mastery of change.
In our time, the pace is quickening.
— Carl Sagan

When Darwin's theory of evolution by natural selection was young, it had a competitor in Lamarckian evolution. Jean Lamarck's theory held that whatever traits an animal develops before it has offspring, it can pass on to those offspring. For instance, a giraffe may stretch its neck every day to reach higher leaves on trees, just as an athlete gets stronger by training. When the giraffe has children, that longer neck trait would be passed along to them. Interesting, but wrong.

Stretching its neck, the giraffe does not alter the DNA code in its sperm or egg (i.e. germ) cells. Since its offspring are products of that code, they will not be affected by how much the parent stretches its neck. So Lamarck, speculating long before Watson and Crick discovered DNA and the underlying mechanisms of evolution, was wide of the mark. What we believe does happen is that random mutation in its genetic code can create a longer neck and that affords the giraffe access to leaves out of reach of its competitors. Better fed and stronger, that giraffe can have more offspring, which share the code for a longer neck. So, the percentage of giraffes in the total population with longer necks increases because that trait has positive selection pressure.

Why dredge up an old evolutionary theory that was wrong? Because it may apply to technology. Retention in technology does not depend on germ cells. Information is retained around the technology as designs, so a design improved over the life of a technology can be incorporated into its successors. That may allow technology to evolve much faster than biological systems, but it raises a question. Why does life not use Lamarckian evolution?

Perhaps it was just chance that biological systems never developed a mechanism to exploit that process. It may require complex systems that are unlikely to arise through random mutation. Or, perhaps, there are disadvantages. If evolution is too fast, can that make for weakness? Is it less risky to allow only slower changes and test them over many generations for fitness before allowing them to crowd out the old, but proven designs? We may find out.

In the meantime, we already know much about how technology changes. While it may follow Lamarckian evolution more than Darwinian, technology does exhibit the three key elements of any kind of evolution: variation, selection, and retention. Memes rather than genes carry the information in this process. Although artificial evolution—the invisible battles of information ecosystems residing in computer memory—may appear but a novelty now, it does employ the powerful forces that created dinosaurs, humans, and AIDS. And it accelerates those processes to evolve thousands, if not millions, of generations each second. The results may not stay hidden inside computers.

Using computers to simulate evolution and develop new technology that we might never think to design is a perfect example of autocatalysis. Computers act on themselves to change themselves, evolving software and hardware so complex that no human team could design it. Then that software and hardware could be used to evolve systems even more complex, and so on. Autocatalysis started simply enough with crude tools allowing humans to fashion finer tools. Since technology has come to extend not only our muscles, but also our brains, the autocatalytic process is freeing itself of human dependence. The programmer may not even understand how the products of simulated evolution work because they are so complex. When humans are not needed even to setup the environment and define the goals, then simulated evolution can accelerate even more.

Can humans keep up with this accelerating rate of change? In the next chapter, we will look at how we have tried, but for now, let's consider a theory that claims we cannot. The many processes that become exponentially faster suggest to some that we are approaching a point at which change will be wholly incomprehensible to humans. They call that point the "Singularity."

Although technological generations lasting just a few years (e.g. computers and cellular phones), have been challenging to many of us, generations lasting hours or even minutes would be too quick for anyone to understand and evaluate. No sooner would we start, but a hundred successive generations would evolve, leaving the one under our consideration extinct or at least obsolete. And exponentially accelerating change would continue past that point, creating new generations of technology we cannot even imagine within fractions of seconds. And then faster.

If the Singularity occurs, we will not be able to understand the details of change, but we will still recognize the forces of change. Technology will still act on itself to change itself. Advantage, compatibility, risk, and visibility will still be selection factors for what survives, though these four factors may be seen through eyes that are more technology than human. Necessity will still create new inventions, and inventions will still create new necessities. Technology will still disappear by being incorporated into other technology and by becoming so common it is no longer noticed (It is not just human psychology that focuses on the novel. Understanding the Universe in complete detail is impossible, so any perceiving organism or system must filter for those details that matter most).

But back to the present. Deciding if we want the Singularity to happen, or if we even have control over the outcome, rests on understanding the forces of change. Unfortunately, few people examine technological change, and when they do, they tend to focus on a particular technology: The computer purchased last year is already slow by current standards; the cellular phone purchased last year is already large by current standards. But technology is like a forest and those particular examples of it, but trees. To see the grand trends, we must uncover the sorts of patterns we did in this chapter. And there is urgency to this. As Carl Sagan pointed out, the pace is quickening.

Chapter 6

How does Technology Change Us?

We are mirrored by our machines…
we mirror our machines.
The question is not whether
we should let them change us,
but whether we are to be lifted up
or dragged down in the process.
– John Lienhard

The pilot of the small propeller plane empties his bottle of Coca Cola and drops it out the window. It whistles briskly down to the barren and unpopulated Kalahari Desert in Africa…unpopulated except for the Bushmen.

These pygmies, with their distinctive language of clicks and whistles nearly irreproducible by Western mouths, find the glass bottle intact and presume it a gift from the gods. The bottle is unlike anything they have ever seen. Clear like water, yet harder than any of their sticks or bones (their territory has no stones), the bottle becomes immediately useful in grinding seeds and grains for food. It also whistles when blown across, helps to stretch hides, and imprints perfectly round circles of colored pigment as decoration on hides. The new technology changes how they prepare food, play, and create art.

But while they have multiple bows, arrows, hides, and other possessions, they have only the one bottle. In a tribe that shares everything, this scarcity inspires competition, jealousy, and violence previously unknown. Convinced that a gift that causes discord must be a mistake, the tribe's leader throws it back to the gods. The gods apparently refuse it, allowing it to fall back to earth, accidentally striking a child on the head. Since the gods won't accept it back, the tribe's leader buries it. Unfortunately, the scents they have left on it attract hyenas that dig it up, leaving it in plain view.

This story of how technology changes us comes from the movie *The Gods Must Be Crazy*. But technology need not be as novel as a Coke bottle in the Kalahari to have profound effect on us. Throughout history, it has changed our lifespan, work, thought process, and the very nature of our species:

- Improving nutrition, enhancing sanitation, and fighting disease, technology has extended human lifespan from 20 to 30 years in prehistory to about 78 in the modern developed world. It has also left us more vulnerable to disease by concentrating us into cities, providing us more calories than we need, and allowing us to become sedentary. Still, the overall trend has been toward longer lives, and that has changed how we lead them.

- The planting stick and the plow changed how we worked to feed ourselves. The surplus of agriculture allowed specialization, leading to occupations beyond hunting, gathering, or farming. Technology has made some jobs obsolete even as it has created new ones. Cars and trucks evaporated demand for blacksmiths and a host of horse-related positions, while employing mechanics, engineers, oil riggers, and car washers.

- Centuries ago, what people considered "fact" was only what they or close friends had witnessed. That changed with the printing of books and newspapers. Now our perception of reality is also influenced by 24-hour television. Through technology we also find out how we are doing and if we should be worried. Drugs such as methamphetamine that simulate the sense of accomplishment can further modify our perception of reality. Future technology may incorporate that mechanism into a dangerously seductive virtual reality.

- In a world governed by "survival of the fittest," technology has changed what "fittest" means and, so, it has affected the evolution of our species. Those who threw spears accurately or worked within the growing social network tended to survive and pass their genes along. Eyeglasses and medical treatments made many physical conditions irrelevant in determining who could propagate their genes. With genetic engineering, changes to our species may no longer be slow. Physically incorporating technology into our bodies may have even greater impact.

Technology is part of our environment, and as our environment changes, so do we. In the previous chapter, we looked at how technology changes as if we were able to stand outside its field of influence. But the observer and the observed are inseparable, and part of our understanding of technology must include how it changes us. We start with investigating lifespan and what technology has done to change it.

> Ever since we crawled out of the ocean
> and stood upright on the land
> There are some things that we just don't understand:
> Relieve all pain and suffering
> and lift us out of the dark
> Turn us all into Methuselah—
> But where we gonna park?
> — Don Henley

Methuselah's Burden

The patriarch Methuselah was reputed to have lived 969 years. If true, that would have been much longer than the 20 to 30 year average in human prehistory. Since then, average lifespan has greatly increased, with technology both helping and hurting its progress. If we continue this trend, we may create a world where Methuselah would be average. Some of the technological effects on human lifespan:

- Agricultural technology (harvesting sickles, scythes, plows, irrigation, crop rotation, chemical fertilizers, selectively-bred plants, pesticides) created denser populations.

- Transportation technology (trucks, trains, ships, airplanes, telecommunication) connected populations, creating a ripe environment for infectious and parasitic diseases.

- We combated those diseases with advances in sanitation and medicine. Water treatment, vaccines, and a host of related technologies greatly increased average lifespan, subjecting us to diseases of old age.

- New technology addresses the degenerative afflictions of the old: circulatory diseases, cancer, diabetes, and the general breakdown and mutation of our cells.

- Technology still in the conceptual stage could augment our natural immune and regeneration systems to the point of giving us near immortality.

Start with agriculture. About 12,000 years ago, humans used sickles made with flint blades and stone handles to harvest naturally occurring grains. Since then, we have made our environment increasingly specialized, changing an ecosystem that happened to support humans into one designed exclusively for feeding them.

The yield of food from a given area shows this progress. Slash and burn agriculture with primitive implements, such as the scythe, produces 225 to 450 kilograms of cereal grain per hectare of land. This approach was typical of Northern Europe during the Bronze and early Iron Age. Add animals for plowing and slaves for weeding and the harvest increases to 500 to 750. Add the heavy plow, which dug deeper than earlier plows (turning the soil over instead of just parting it), to produce 600 to 900.

And when you have soil so light that even a curved stick will turn it (such as irrigated desert agriculture), that produces 1200 or more kilograms of cereal grain per hectare. Sometimes much more: the Nile River basin produced as much as 2500 kilograms of cereal grain per hectare of land during Roman control. The key there

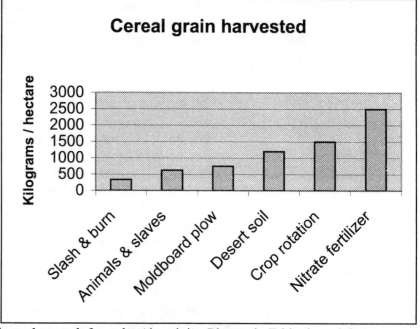

was fertilizer: the Nile brought potash from the Abyssinian Plateau in Ethiopia and decomposing vegetation from Lake Victoria all the way down and through Lake Nasser.

Where soil had no such natural mechanism for fertilization, people developed techniques such as crop rotation. Planting nitrogen-fixing legumes after a grain harvest restored the soil so that it could support another planting of grain. In England this produced 1500 kilograms of cereal grain per hectare of land.

To do better than that required nitrate fertilizers, which allow a hectare to produce 2000 to 3000. Breeding of new varieties of cereal crops that can absorb more fertilizer (without becoming over stimulated) can produce 3000 to 6000 and more.

Specialization of ecosystems in the east paralleled the west. In Japan, rice production advanced from less than 1300 kilograms per hectare under primitive farming to well over 5000 using irrigation and other technical innovations of the present day. And now the world is one, with pesticides and genetically modified crops available—if not necessarily affordable—to all. Boats, airplanes, trucks, and trains move these agricultural technologies as well as the foods they help produce around the globe. Between hothouses and the global transport network, fresh fruits and vegetables are available year round.

Increasing food production from several hundred kilograms per hectare to more than 30 times that has led to population increases of even greater proportions. Population not only increased, but also became much denser because farming surplus allowed people to congregate in cities, specializing in non-farming activity.

Thousands of years ago, when all this was first happening, we could not have known that urban concentrations would be attractive environments for infectious disease and, in particular, epidemics. Now we know three reasons they are: *sanitation, density,* and *size.*

1. For most of the time we have had cities, they have been unsanitary, with garbage and sewage spilling out onto streets and waterways. Although domestic bathroom plumbing first appeared 4000 years ago (on the island of Crete) and the modern flush toilet was invented in 1884, major cities today continue to dump untreated sewage. For instance, Karachi, a Pakistani city of 15 million, releases sewage directly into the Lyari River and most of its residents lack access to clean water. Cholera and other diseases spread through fecal water contamination.

2. Disease loves cities because new hosts are packed close together. Whether transmitted through direct contact or intermediaries like rats, fleas, or contaminated water, it is easier to find someone new to infect than it would be in sparsely populated rural areas.

3. Epidemic diseases such as smallpox, measles, cholera, and influenza leave their hosts either immune or dead. These diseases are on "sinking ships" and must find fresh hosts to survive. Fortunately for them—and unfortunately for us—cities support large populations with enough newborns and immigrants to keep all but the most deadly epidemics circulating.

A devastating example of epidemic diseases requiring a minimum population comes from the Faeroes, a small group of islands between Scotland and Iceland. The islands suffered a measles epidemic in 1781, but, with fewer than 10,000 inhabitants, they could not support that disease. With many dead and all the survivors immune, measles died out, disappearing for 65 years until a visitor from mainland Europe reintroduced it.

Some diseases are unable to survive in even larger populations. A direct overland trade route between China and Europe brought, among many other things, bubonic plague or "Black Death." Killing one in four, it burned out even among a population of millions. And, by the end of World War I, the world was so well connected that an influenza pandemic killed 21 million before it ran out of victims. In the 21st century, transportation technology connects almost seven billion people into a community of hosts large enough for epidemics even more deadly than that. People now circle the world in much less time than it takes for most diseases to kill their host...or even display symptoms.

In the chapter on where technology comes from, we noted the great benefit of concentrated (and, by the same argument, connected) populations in allowing one innovation to trigger another. Like disease, ideas thrive in environments where many hosts can share them.

The people suffering in filthy, disease-ridden cities shared ideas about improving the situation. Two of the most significant areas of improvement were in sanitation and medicine. Water wells, chlorination, and waste treatment have enhanced sanitation, protecting us from water-borne diseases like cholera; sanitation in food-handling practices protects us from food-borne bacteria like salmonella. Antiseptic techniques, quarantine, antibiotics, vaccines, public education, and the sharing of medical knowledge have also extended our average lifespan.

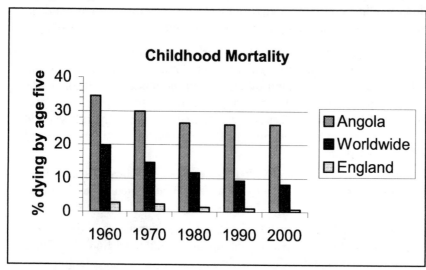

And human lifespan has extended remarkably. As we noted earlier, thousands of years ago, human lifespan was between 20 and 30 years. Between 1800 and 1900 in England, it climbed from the mid-30s to about 50, and then beyond 78 by 2004. A major factor has been childhood mortality—dying before the age of five. In prehistory it was as high as 50%. Even if the surviving 50% had all lived to 50, which they did not, average lifespan would have been little more than 25. Worldwide childhood mortality dropped to 19.7% by 1960, 8.3% by 2000, and 8.1% by 2002. Angola and England continue to bracket this, holding steady over recent years at 26.0% and 0.7% respectively.

Child mortality does not affect everyone equally, but tends to weed out those who are weakest. When it is high, mostly the strongest survive to old age. When technological advances cause childhood mortality to fall, we might expect the swelling ranks of older people to include many that are so weak that they succumb to other ailments. And, yet, life expectancy at older ages has increased where child mortality has dropped. In the U.S. a 65-year-old living in 1960 could have expected another 14.3 years, and one living in 2000, another 17.9 years (this is different from life expectancy *at birth*, which is now 77 in the U.S.). So, while the largest improvement to average lifespan has come from reduced child mortality, older people are living longer, too.

There is much room for progress in improving sanitation and medicine around the world. In 2002, 1,798,000 people died from diarrhea, 1,566,000 from tuberculosis, and 1,272,000 from malaria. Where modern sanitation or medicine is absent, lifespan is short (e.g. less than 37 years in Angola and 36 years in Zambia), largely due to absence of these technologies and presence of others. Weapons technology has made it easier to wage civil wars, which shorten lifespan both directly and by preventing development of infrastructure to deliver clean water, safe food, administration of vaccines, and basic medicine. Conflict also displaces people, and refugees can neither farm nor readily conduct commerce.

In the developed world, free of civil war and generally replete with food, medicine, and basic sanitation, lifespan faces new challenges. More food does not necessarily lead to longer life. Not only can we consume calories with little nutrition, we can consume so many calories that we become vulnerable to diseases of overnutrition. Obesity contributes to both circulatory disease and diabetes. So while our technology can provide almost any foods, many of those in a position to choose, opt for fried animal products high in saturated fat and processed foods high in refined sugar. Such lifestyle choices reduce average lifespan, though a variety of drugs and interventions can keep us alive, if neither healthy nor independent. Perhaps surprisingly, some of the longest living individuals subsist on low-calorie diets.

For those who make the healthiest lifestyle choices there are still limits to lifespan. Genetic imperfections can lead to a variety of ailments, including heart disease. And even without genetic problems from birth, each year of life exposes us to genetic mutations. These come from ultraviolet sunlight (melanoma), radiation and chemicals (leukemia), and time (various cancers and old age). The human body has 50 to 75 trillion cells, each reproducing as often as hours (white blood cells live 10 hours) or days (skin cells live 19-34 days). There is always the chance of DNA copying errors when cells divide, so the longer we live, the more mistakes we contain. Usually, the body detects these mutations and destroys them, but there is always the chance that some will be missed and become cancerous.

Time may also play a more basic role. Some human cells divide only a certain number of times, with a telomere sequence at the end of a chromosome counting each divide. On average, the skin cells of younger people have longer telomere sequences than those of older. This mechanism may help prevent the runaway divisions found in cancer, whose cells divide without limit, but it may also impose a limit on human lifespan. Whatever further research concludes about the role of telomeres, the human body may have various forms of built-in counters, limiting the lifespan of even a perfectly nourished, disease-free human.

This means that technological progress in nutrition, sanitation, and disease can help to spread our "best practices" to the billions still wanting, but may do little for those already benefiting from them. However, advances in emerging areas, such as genetic engineering and nanotechnology, may break this lifespan ceiling.

If nanotechnology develops to the point where robots just hundreds of nanometers on a side (nanobots) could patrol our bloodstreams, then that might lead to the next leap forward in lifespan. Imagine millions of submarines, each smaller than a blood cell, programmed to identify unhealthy cells and destroy them. Our natural immune system is remarkably resilient, developing new antibodies when exposed to new threats, but the environment in which it evolved has changed and continues to change at an accelerating rate. Nanobots could help us keep up, downloading distinguishing characteristics of the latest viruses or forms of cancer and protecting us from afflictions even before exposure.

Our individual immune systems would, in effect, be networked globally. The World Health Organization, the U.S. Center for Disease Control (CDC), or some similar entity, could distribute "fingerprints" of viruses, cancers, or new bioweapons as soon as they have been identified. Though to further extend lifespan, it may be insufficient to eradicate the dangerous elements from our bodies. We may have to develop ways to promote healthy development of replacement cells.

Chess-Playing Computers and Bacteria

A computer programmed to play chess illustrates the power of being able to share what we learn. Although it may take a long time to program a computer to play grandmaster-level chess (just as it takes human grandmasters a long time to learn), that program can then be copied to millions of other computers. Once a process is defined, refined, and automated, the algorithm to perform it is replicable and shareable. It is just information.

By contrast our biological immune systems learn, but cannot readily share that learning with the immune systems of other people (the exceptional case: a mother does confer temporary immunity on her infant). Imagine if whatever one person's immune system learned could be shared with millions of others. As fantastic as it sounds, the sharing of biological immunity happens today, though only between bacteria.

Because they do not shield their genes within a double-walled nucleus, bacteria can share strings of their genome with each other. Once a bacterium develops resistance to an antibiotic (perhaps by random mutation), it may share the genetic plan for that resistance with its neighbors, and they with their neighbors. Bacteria that incorporate something as useful as resistance to penicillin, are more likely to live, multiply, and share that genetic information. Similarly, those incorporating genetic information harmful to themselves are more likely to die and be unable to share it with others.

Plants and animals, including humans, cannot do this naturally because our cells, called eukaryotes, do shield their genes within a double-walled nucleus. Eukaryotes evolved from prokaryotes (e.g. bacteria cells) about 1.5 billion years ago. Technology may bring back to us an ability to share immunity that we lost long ago—this time operating not just locally, but globally, and at the speed of light.

Rather than begin a detailed technical analysis of how nanobots could function or worry about how that technology could go horribly amiss, let's consider the implications *if* nanobots or some other technology could greatly extend lifespan. Over a long period of time, lifespan has trended ever higher, and there is little reason to believe that would suddenly stop.

So, suppose some humans started living far longer than any do now. Human culture and mythology are aligned with limited life. Although lifespan has more than doubled from less than 30 years in prehistory, this has been a very slow process and we have had time to become accustomed to patterns of childhood, adolescence, discovery, settling down, reevaluating life, creating a legacy, and reconciling with death.

We have already noticed some changes in less than the past century. People are living well beyond the usual age for retirement from work. When lifespan averaged 60, many died shortly after they left their jobs. Now, many are reinventing themselves, going back to school or starting entirely new careers or new families. If some technology makes human life indefinite, how will people cope? Will they exploit the ever-growing percentage of life that can be directed by experience and wisdom—avoiding the joke "I just figured it out and now I'm too old to do anything about it"? Or will indefinite life sap the urgency to seize the day?

The more important question to some: will we be healthy to enjoy those extra years? The "opportunity" to live an extra few hundred years connected to life support machinery is appealing only to those in abject terror of death. Independence and vitality until we choose to conclude our lives may be the only conditions under which many would opt for far longer lives.

The longer people live, the longer they tend to wait before having children and the fewer children they have. The highest birthrates are in the poorest countries, which also have the highest infant and overall mortality rates. Birthrates in some developed countries are insufficient to balance death rates. Italy averaged 8.9 births and 10.1 deaths per 1000 people in 2002. Were it not for immigration, Italy's population would shrink. Would near-immortality cause population to explode? Or would decreased death rates further depress birthrates, keeping a balance?

How would limitless lifespan affect education and work? How would it affect our systems of government? Or family relationships, which could grow far beyond three or four contemporary generations? How would 20-year-olds and 700-year-olds interact when they appear the same age? Would society segregate by age so that experienced and cunning 500-year-olds do not prey on the young and innocent? Would we find that nearly everyone who has lived for centuries loses interest in those who have not? After all, they may have little in common. Or, would wildly unpredictable reactions come of such diversity?

Although homicides and suicides are committed for a variety of reasons, rates of incidence have followed long term trends along with health and longevity. In the year 1300, England had nearly 20 homicides but less than one suicide for every 100,000 inhabitants. Homicide rates dropped and suicide rates increased so that around 1700 they crossed at about five per year. By the year 2000 in the United Kingdom, homicides fell to 1.3 and suicides climbed to 8.7. Would an immortal be much more likely to commit suicide than kill someone?

Longevity affects us in many ways. Is it possible to live 1000 years without perceiving the Universe differently? Until recently, such thought games have been of use only in intellectual circles to probe abstract ethical and moral issues. But with some technologies changing at accelerating rates, as we noted in the previous chapter, we may live to see our questions answered.

> We are brothers and sisters of our machines.
> Minds and tools have been sharpened against each other
> ever since a scavenger's stone fractured cleanly
> and the first cutting edge was held in a hunter's hand.
> — George Dyson

Working up the Pyramid

Stone tools, spears, and bows & arrows changed how we hunted. Scythes and plows turned hunters into farmers. Clocks regimented our work. Steam engines and automation turned farmers into factory workers. Ships, airplanes, trains, trucks, and telecommunication systems moved factory workers from one country to another. And they still do, creating and destroying jobs in each. As Nietzsche observes, these technologies have a vast and rippling impact.

These changes can be disruptive. Archeological evidence shows that the first farmers were less healthy than their hunter-gatherer relatives. Bones of those early adopters of agricultural technology exhibit arthritis, rickets, tuberculosis, and other deformities. So, while agriculture produced surplus to get everyone through hard times (when hunters would simply die off), farming was harder work. It succeeded in spite of this, perhaps because 10 to 100 times as many farmers or herders could survive in a given area, giving agricultural societies the force of numbers to conquer and displace.

In 19th century England, technology dramatically disrupted how we worked. Automated factories displaced the cottage industry of textiles, efficiently producing more with less labor. Faced with starvation for themselves and their families, desperate workers smashed the machines that took their jobs. Naming themselves followers of the mythical Ned Ludd, they were the first Luddites—a term that now describes anyone who views technological progress as dangerous and harmful. The Luddite protests were violently suppressed and the machines rebuilt before eventually giving way to even newer machines and newer forms of work.

Some cultures have successfully banned such disruptive change. Amish communities in the U.S. accept 19th century technology as part of their normal lives, but resist anything newer. Others draw a more flexible line to identify desirable changes to how we work. Mohandas Gandhi used a spinning wheel and advocated its widespread use in India in order to achieve independence from England in production of cotton cloth. Economist E.F. Schumacher described technologies appropriate to underdeveloped areas in his book *Small is Beautiful*, identifying non-economic measures of success, much as Gandhi had. With this approach, a technology that was affordable, operable, and maintainable by the indigenous population would be evaluated as better than one that simply provided a higher return on investment.

These are serious choices since technology has huge impact on how we work. But what are the patterns of that impact? Technology makes us more *specialized*. It replaces *brute strength*. And it both creates and replaces *repetition* in work.

Specialization - Agricultural technology enabled farmers to grow more food than they needed for themselves, so they could trade surplus for goods made by non-farmers. It has reduced the U.S. farm population from about 95% of the total in the 18th century to less than 2% today, with the rest involved in a complex web of interchange. Technology allows us to create value in this web (and therefore survive) doing ever more specialized tasks, ever farther removed from making food. The printing press made books inexpensive, which connected geographically-disperse people with similar interests. While letters had allowed one-to-one connections, books created intellectual communities fostering groups with specialized knowledge, vocabulary, and work (such as doctors, lawyers, engineers, and scientists). And newer communication technologies, such as the Internet, continue this process.

Brute strength - Technology can provide superhuman power. Outside of carefully regulated sporting rituals (such as football) our work depends less and less on strength. We still have ditch-diggers, but they do the finer work for which bulldozers, trucks, and cranes are still too crude. We still have assembly-line workers, but robotics has slowly started to displace them, too. Increasing robotic sophistication and decreasing technology costs will continue this trend, forcing us toward work that demands ability beyond brute strength. That strength need not be just physical. Brute intellectual strength could describe a computer's ability to perform billions of calculations—a lifetime's work for a human—in each second. By contrast, we humans still need to perform more delicate feats of intellect, such as expanding the frontiers of science.

Repetition - Repetitive tasks are easier for technology to perform than are tasks that vary depending on circumstance. Humans evolved to be adaptive, surviving in both Siberia and the Sahara. Technology has not been so versatile, but it has been very effective at repeatedly doing the few things it was designed to do. Assembly lines show this well, with each station in a long line performing one task like clockwork. Ironically, even as technology appropriates repetitive tasks, it has often pushed human assistants into pacing it with similarly repetitive work. The work that the Luddites wanted back (so they could buy food) was repetitive combing and spinning of wool on simple machines in the home or shop. The work that displaced them was repetitive setting up and feeding of steam-driven machines in factories.

Though technology in Third World sweatshops still demands the repetitive assistance of many people, the trend appears to be toward technology that rarely needs us. Robotics has led to factory machines that can run with little human intervention. The repetitive elements have evolved into computer programs running again and again. In the Deep Blue computer program that repetition was sophisticated enough to beat world chess champion Gary Kasparov. Deep Blue tested 200 million board positions each second—perfect for a game with explicit rules—but it could not have matched Kasparov's navigational ability in getting from airplane to taxi cab to hotel and chess venue. Since humans tend to have broad, contextual knowledge, technology still relies on us when circumstances change.

Where do these patterns of specialization, brute strength, and repetition point? Some guess that humans are being pushed up a pyramid of work by technology. At the bottom of the pyramid is work that is concrete, algorithmic, repetitive, and isolated. At the top, it is abstract, creative, novel, and contextual. As technology advances, it becomes capable of work higher up the pyramid. As if retreating from rising floodwaters, humans will continue to work at higher and higher levels.

Although highly sophisticated, Deep Blue's chess playing ability fits at the bottom of the pyramid. The objective is *concrete*: capture the opponent's king. The approach is nearly *algorithmic*: apply a set of rules to evaluate various board positions that tends to indicate a path toward winning (technically, algorithmic would *always* find the path, so this is more heuristic than algorithmic, placing it a bit up from the bottom of the pyramid). The technique is *repetitive*: the same rules are applied millions of times each second to evaluate billions of possible moves. The scope is *isolated*: the rules and objective of chess are unaffected by context, such as the weather, politics, or the personal history of your opponent (Kasparov's travel to the match, by contrast, depended on many environmental or contextual factors).

Albert Einstein's development of the theory of relativity is work near the top of the pyramid. His objective was *abstract*: he sought some model for the Universe that resolved contradictions in the previous model. His approach was *creative*: he imagined what it would be like to ride a trolley car at the speed of light, which is hardly an obvious thing to do. His technique was *novel*: there is no recipe for developing new scientific theories. His scope was *contextual*: any model for how the Universe works must satisfy and agree with a range of existing observations and theories. How did Einstein know whether his theory must agree with what was known or speculated in the fields of physics, chemistry, biology, or information theory? His sense of the scope of the task might be called intuition, placing it near the top of the pyramid.

Some might dispute that the technology tide is rising on the pyramid, pointing out that many people in the developing world still farm the land or work in factories making sneakers or assembling computers. True, but the rising tide describes the most advanced capability of technology, not the degree to which that capability has diffused into widespread use. Diffusion depends on political, social, economic, and cultural factors, so it can be a slow process. And it can also be a quick one because technology replicates easily. For instance, software developed over years to accomplish a specific task can, once finished, be copied around the world in just minutes.

Ultimately, this model raises two questions. Will the pyramid of work continue to expand upward, presenting ever more abstract, creative, novel, and contextual challenges? And, if so, will humans continue to scale the pyramid, staying a few steps ahead of the most intelligent machines?

The answer to the first question is probably "Yes" because history shows we have always found more challenging tasks every time technology freed us from routine drudgery. For instance, prior to electronic computers, all calculations were performed by hand. People (whose job description was "computer") figured taxes, ballistics for shooting cannons, engineering problems, and everything else that required the manipulation of numbers. With the advent of electronic computers, capable of performing more calculations in a single second than a human could in a lifetime, people started programming computers instead of computing numbers.

That leaves the second question. If humans stay ahead of technological capability, we may experience an environment similar to that shown in Star Trek, with computers supporting us as we seek out and explore new worlds. If not, technology may develop human intelligence and flexibility, and then continue past us.

> Many of the products of this accelerating process—written language, city-states, and automation, for instance—sped it further. Today the pace strains the limits of human adaptability...Nevertheless, the acceleration continues, as machines take over where humans falter.
> – Hans Moravec

Our Perception of Reality

Hunting down dinner or finding some nutritious roots triggers an electrochemical mechanism in our brain. Release of the chemical dopamine (also triggered by getting a promotion, assembling a piece of furniture, or receiving applause) makes us feel euphoric. Natural craving to re-experience this euphoria motivates us to further accomplishment. Technology—since the first tools—has helped us accomplish goals and bring on this feeling we so desire. But it is also helping us short-circuit this motivation mechanism.

By stimulating release of dopamine, a drug like methamphetamine simulates the effect of accomplishment. It gives us the payoff with none of the effort. Unable to precisely control timing and effect, methamphetamine is a crude tool to change our perception of reality. Through sight and sound, video games create a different kind of reality. They entertain us and give us a sense of accomplishment, whether from wiping out aliens, scoring a touchdown, or racing a car. Now, imagine combining the two: a video game that could precisely stimulate dopamine receptors. If it also made us see and hear a fully realistic world, how many could resist?

Pieces of this compelling virtual reality are already here. In 2002, a team led by bioengineer Sanjiv Talwar attached electrodes to three areas in the brain of a rat. One stimulated a pleasure center—an electronic equivalent of eating or drinking—and the other two simulated sensation of the rat's left and right whiskers. The "pleasure" electrode was used to train the rat to turn left or right based on whether the "left" or "right" electrodes were activated. Once trained, the rat could rapidly navigate a maze under human control. This is still a crude control, but more recent experiments have connected to hundreds of neurons in the brains of monkeys, and newer technologies promise non-invasive interaction between brain and technology.

Nanobots injected into the bloodstream could follow the same circulatory path that drugs take to the brain. Designed to attach to specific parts of the brain, such as dopamine receptors, they would "listen" for wireless instructions from the video game console. If the nanobots were designed to stimulate other parts of our brain, they could also simulate sight, sound, and touch. Imagine skiing down Mount Everest, seeing the crystal-clear glacier fields rushing up, hearing the crunch of snow and ice, feeling the cold wind and the push of the skis against your feet...and the dopamine rush of doing what no human could really do.

Since the relatively crude drugs we have today distract many addicts from working and even eating and sleeping, just imagine the impact of an experience far more immersive. Will our principles compel us to give up the sensation of anything we desire? If our brains were experiencing every dream we could imagine, would we care that our bodies were slumped in dark rooms? This is a slippery slope: Once we start, what could motivate us to stop short of the abyss? Simply knowing how to use such a technology—rather than understanding and evaluating it first—would be dangerous.

Of course such technology could also benefit us by furthering education and exploration. Imagine hovering above the earth and watching millions of years of tectonic movement in seconds. At will, you could zoom down into the earth to view, hear, smell, and feel the underlying processes, even microscopic interactions of different materials. Because it would all be computer-generated, you could change any of the variables at any time. Such sensory immersion would provide an unparalleled intuitive sense of the theory and mathematics behind complex processes, making them easier to study and understand.

Even though current technology cannot create this level of virtual reality, technology has long been changing our perception of reality. The earth was widely accepted to be the center of the Universe until 1609, when Galileo used the newly invented telescope to view moons circling the planet Jupiter. Church doctrine maintained that everything in the skies circled the unmoving earth. Since Jupiter's moons appeared to circle something other than the earth, perhaps the entire doctrine was flawed. What a change in reality that must have been: the seemingly solid and unmoving earth just another bit of rock spinning around the sun. Since then, newer astronomical technology has relegated our sun to one arm in a spiral galaxy among many galaxies.

In the other direction, microscopes revealed an invisible world all around us. They revealed the microorganisms constantly crawling all over our skin and the living cells that compose that skin. The microscope led to germ theory, which explained the source of infections. In this new understanding of the world around us, doctors began taking the novel step of washing their hands and instruments before operating. Now we know that many deaths ascribed to other causes were due to the septic medical interventions intended to save lives. What will future generations know about the causes of disease and death that we do not? What technologies will change their perception of reality to include that which is still invisible to us?

Just as telescope and microscope have brought us new information, so has communication technology. What people knew 600 years ago was what they saw or what a trusted family member or friend told them they saw. Many lived their entire lives in villages, so their perceptions of the world were based on oral myth and folklore. And then Johannes Gutenberg invented moveable type. Books, newspapers, and other printed material brought information from afar, painting a different—we would say more complete—picture.

Radio, television, and satellite technology made this information not only more complete, but more current. It took two months after the fall of Constantinople in 1453 for word to reach Rome. Today we would know within hours, if not minutes, by watching CNN's or Fox's 24-hour news. The Internet offers alternate and contradictory views just as quickly. Since a complete picture of the entire world is far beyond anyone's ability to absorb, the flood of information is filtered. And that distorts reality in a new way.

Take, for instance, the apparent phenomenon that disasters often come in groups. After three airplanes crashed into buildings on September 11, 2001, many more cases of plane crashes—some into buildings—were reported. Was it some strange coincidence, were they further terrorist acts, or were these copycat acts? Actually, the number of airplane crashes was typical, but after the drama of jets hitting the World Trade Center any sort of airplane crash became newsworthy.

The same happens after a particularly dramatic instance of school violence or food poisoning: news that would have been lost in the background gains exposure. Of course, there may also be imitative acts, and these will be particularly well covered because they fit a pattern. Because our media technology can provide far more information than anyone can digest, those in control of that technology make choices about what will be communicated. And that creates our reality.

Perhaps because of a psychology evolved in a much simpler world, many find comfort even in simplifying prejudices: good vs. evil, our religion vs. theirs, our ethnicity vs. theirs. Technology mediates between our environment and us, so it can reinforce the perception of any reality that we want. If we want to believe that the CIA or Mossad orchestrated the flying of planes into the World Trade Center, we can find websites that document and confirm this. Technology channels ever more information to us, leaving us to our traditional techniques for filtering out the truth. Increasingly, power exceeds wisdom.

From telescopes, microscopes, and printing presses to the global telecommunication network and the futuristic virtual reality described at the beginning of this section, technology changes how we comprehend our world. At some point in history, it must have seemed that technological progress would take us ever closer to "true" reality. Scientific tools gave ever more accurate pictures of how the Universe operated. Information flowed more easily and quickly around the world. Even if a perfect grasp of reality were not possible, we were making constant progress.

But that may not be the case. Our psychology limits us and our physiology makes us vulnerable. How much information can we absorb? How tempted are we by simplifying prejudices? Do we really want to achieve or are we satisfied with the sensation of achievement?

Evolution does not create from scratch, but builds layer on layer, so humans bring baggage from an environment quite different from the current. We are the genetic legacy of those who were fittest to survive in a world before methamphetamine, virtual reality, and the information overload from books, media, and a variety of technologies. The psychological and physiological products of our history combine with continually changing technology to form our perception of reality.

Our legacy of psychology and physiology is not immutable, though. Technology not only interacts with it, but also changes it. It has already changed our species and, as we have observed, that change is accelerating.

> If we are to maintain a primary role on this planet,
> we must understand technology and use it to advance our own
> limited brain capacity. It is not possible to ignore these changes
> for they are inexorable, and will promote even more change.
> In short, you can opt out, but you cannot escape!
> – Peter Cochrane

The Human Species

Since it was first invented, technology has been changing the human species. From early on, throwing rocks or spears accurately was an important survival skill. Those able to propagate their genes probably had good eye-hand coordination, depth perception, and the mental ability to estimate where a running animal would be by the time a projectile reached it. Primitive technology selected for certain traits in us, influencing human evolution just as changing environments have influenced the evolution of all life.

Trains, ships, and other transportation technologies allowed people to travel farther in finding mates—rather than marrying "the girl next door" or the boy chosen by the village matchmaker. This creates greater diversity in the genetic makeup of children.

The human gene pool has been further affected by technologies that make otherwise fatal defects irrelevant to survival and procreation. Eyeglasses let the myopic see approaching threats. Blood-glucose monitors and insulin allow diabetics to survive. Of subtler influence than these obvious corrective technologies are ubiquitous technologies, such as electric lights, which allow us to ignore the cycles of night and day in which we evolved.

When environmental selection factors change, so does the species. Take the dog, for instance. Mitochondrial DNA analysis suggests that all dogs evolved from wolves as long as 135,000 years ago. Humans domesticated some, changing their environment. Dogs that pleased us—whether through loyalty, non-threatening appearance, or mild manners—were allowed to stay in our "pack." Since wolves have difficulty surviving outside of a pack, the dogs that we cast out may not have survived to pass on their genes. Through selective breeding, choosing exactly which animals have offspring, we created a fantastic variety of domesticated dogs, from Chihuahuas to Saint Bernard.

While our increasingly technological environment does select for new and different traits in us, most resist, on ethical grounds, selective breeding of humans. It has occurred though. Infanticide of daughters in China has skewed the population to be more male. In Nazi Germany, individuals matching an Aryan ideal were encouraged to mate.

However, technology in the area of diagnostic medicine may popularize selection. Parents can already detect certain genetic defects in unborn children, and may choose to abort a fetus with a debilitating or terminal condition. As biotechnology reveals more relationships between our genetic code and both ailments and aptitudes, there will be more opportunities to intervene.

Selection, whether by a changing environment or consciously directed by humans, has one limitation. It can select only from naturally occurring traits. For instance, if some dogs have particularly forward-looking eyes, they can be allowed to breed, encouraging this trait. This likely happened over a long period of time because the eyes of dogs tend to be on the front of their heads, while wolves' eyes are more on the sides. However, dogs could not be selectively bred to lay eggs because that is not a naturally occurring trait. Technology is removing that limitation.

With genetic engineering, we need not wait for naturally occurring traits. We can either design them, as in gene therapy, or appropriate them from one species to another, as in the case of splicing human genes into pigs for growth of transplantable organs. Much has been written on the potential impact of genetic technology, so in this book we will focus on a different way that technology may change our species, one that could have a far more dramatic impact.

> I think humans will accumulate
> artificial and mechanical abilities,
> while machines will accumulate
> biological intelligence.
> This will make the confrontation
> between the two even less decisive
> and less morally clear than it is now.
> — Mark Pauline

Humans are incorporating technology. Pacemakers correct for irregular heartbeat. Computer implants in the human brain already correct for hearing loss (cochlear damage in the ear) and for some of the effects of Parkinson's Disease.

Progress will be in two directions. First, new implants will be invented to treat other medical problems and, second, implants will be developed, not for therapy, but for enhancement. It is this second thread, along which technology confers new ability rather than restoring lost ability, that we pursue.

Technology has long augmented our abilities, but it has remained external to our bodies. Before the development of writing, knowledge was memorized in the form of rhyme, a practice that continued until the printing press made the written word easily copied and widely accessible. Books liberated us from having to memorize all the information we wished to know. Computers can do that plus index and retrieve libraries in an instant. With an Internet connection (at home, in a public library, or in an overseas café), one can scour the entire World Wide Web in less than a second. To someone living several millennia ago, this publicly available power would be magic.

> Rather than having to carry the factual
> baggage around in your head,
> you have this electronic prosthesis
> — Sven Birkerts

For many, the term "electronic prosthesis" evokes the image of the omnipresent cellular phone. While some have become dependent on computers for keeping track of information, many more are dependent on cell phones for communication and entertainment. How many will be willing to have these external technological enhancements implanted?

Although many might be repulsed at the idea initially, consider how quickly social acceptance came to cell phones. Younger generations are willing to do many things that the older resist. For example, a variety of body piercing—at one time, proper women did not even pierce their ears.

Once advances make implantable technology feasible, safety, and inexpensive, it will likely change from exotic oddity to necessity. Rebellion will be the first motivation and those acting to defy convention will not even wait for safety concerns to be addressed. Competition will be the second motivation, with early adopters undeterred by expense, just as globetrotting executives pioneered the use of early satellite and car telephones. Once the costs drop and "everyone is doing it," convenience and fashion trends will likely motivate the masses to join the rebels and early adopters.

If technology implants confer superhuman memory, calculation, searching, and communication, then non-augmented humans would be at a disadvantage. This is looking beyond the first step of simply incorporating the functionality available to us in cell phones, personal digital assistants, and laptop computers. This would come only with future breakthroughs allowing synergy between brain and technology.

If such an organic-silicon system could be integrated, then human thought processing ability will hitch a ride on the process that doubles in performance every year or two, a process that has already given us $1000 computers that perform a billion calculations each second and can store a library of information.

Augmented humans would be able to understand complex systems in ways that neither computers nor humans do today. Computers can keep track of millions of details (for example predicting traffic patterns by simulating how hundreds of thousands of vehicles, each acting according to simple rules, would interact). Since humans grasp context (e.g. predicting that traffic will be much worse before and after a major sporting event), a human-computer hybrid may be able to combine both into completely new abilities—perhaps a comprehension of how social systems, ecosystems, or physical systems work, without the many simplifying assumptions on which humans now rely.

Such a future hybrid would probably have as much success explaining advanced concepts to humans as a human might have explaining them to a pet dog. In each case, the human and the dog might sense the general emotion, but not fathom the concepts. Even if non-augmented humans could initially understand the concepts so simple and obvious to hybrids, the performance doubling that the hybrid's technology aspects experienced every year or two would rapidly change that.

The next step would be plugging into a vast network—like a future version of the World Wide Web or, more darkly, Star Trek's Borg collective. Although hybrids would still have human DNA (as well as comparable coding to define the configuration of their technology aspects), they might not be considered the same species.

The difference between a hybrid and a human may be similar to—but far greater—than the difference between our Cro Magnon ancestors and the doomed Neanderthals. Until

30,000 years ago, both human-like species coexisted in Europe. Neanderthals were sturdy and self-sufficient, living in groups of 10 or so. Then came Cro-Magnon with improved communication ability, group size closer to 100, and more advanced technology. Neanderthals, highly capable but a little less efficient at competing for resources, went extinct.

Humans have evolved with technology because we made it part of our environment. Those best able to exploit technology survived and thrived. Likewise, technology best suited to the environment was copied and refined. Humans and technology cooperate, compete, evolve, and may one day converge.

This is history's drift:
technology correlating the fortunes
of ever-more-distant people,
enmeshing humanity
in a web of shared fate.
– Robert Wright

From the beginning, technology changed us, but the effects were initially modest and imperceptibly gradual. Marshall McLuhan's statement "First we build the tools, then they build us" is now coming clearer in many dimensions. Technology directly and indirectly heals us and harms us. It destroys industries and leads to wholly new vocations. It creates and distorts what we believe about our neighborhood and Universe. It changes our very species and may incorporate itself into us to evolve an entirely new species.

Our technological choices change people far away and people yet to be born. When we view technology myopically, our impact can appear minor, even insignificant. But myriad choices have acted over millions of years to transform us from animals we would barely recognize as human. We made stone tools and they made us. And those choices are now coming every quicker and with greater impact. Where do we want to go from here? What do we want to be? What does everyone else want to be? And do we really have a choice?

While some claim that technology is deterministic—that the invention and the impact of certain technologies is inevitable—we have seen examples of superior technology, such as the Dvorak keyboard, being thwarted by human activity. There is much we can do in a variety of roles to guide our creations. Part of understanding technology is recognizing how we can change it.

Chapter 7

How Do We Change Technology?

Build a better mousetrap and the world
will beat a path to your door.
– Ralph Waldo Emerson

The poet Emerson fell for the myth Thomas Edison created. A shrewd promoter, Edison advertised his phonograph, light bulb, and other inventions as fruits of a tireless, lone inventor, but in the laboratory he had a team of specialists looking for market opportunities and then cooperating to address them. And outside of the laboratory, he had investors, distributors, and media contacts. Far from waiting for the world to beat a path to his laboratory door, he promoted inventions before they were even ready.

On September 15, 1878, Edison confidently told reporters that he had invented the light bulb and all that remained was to work out a few details. He predicted that, within six weeks, gaslight would be obsolete. This report drove down the stock of the gas companies and drove up investment in his company. Although those few details took more than two years to work out, he created another media event on New Year's Eve, 1880, when people traveled from miles around to his laboratory in Menlo Park, New Jersey, to see the first enduring light bulbs (he and others before him had made many bulbs that quickly burned out).

Edison was even more creative when alternating current (AC) from rivals Westinghouse and Nikola Tesla threatened his direct current system (DC). By publicly electrocuting dogs and cats with AC, he tried to sway opinion on its safety. Hardly the lone inventor waiting at home.

Why does the myth persist? Because we like simplicity. Just as grand military exploits involving millions of people are credited to individual such as Alexander the Great and Genghis Khan, grand technological inventions are credited to individuals such as James Watt and Thomas Edison. It is easy to imagine one person responsible for one achievement, and marketing and sales people fuel this view because the last thing they want associated with their products is complexity—that could confuse the consumer. Unfortunately, the myth of the lone inventor keeps us ignorant of the variety of roles we can play in the birth and life of technology.

Consider some of the roles played in developing one of the first personal computers. Steve Wozniak, a clever electronic circuit designer and hobbyist, *engineered* the first Apple computer. Steve Jobs *promoted* it, convincing Mike Markkula to *invest* in it so they could produce it on a mass scale. Jobs also convinced John Sculley to leave the top post at Pepsi to *manage* Apple (famously asking him, "Do you want to spend the rest of your life selling sugared water or do you want a chance to change the world?").

Some claimed that IBM's less elegant PC overtook Apple simply because the force of the world's largest computer company made it successful. But there was more to the PC's triumph than that. IBM made public the technical specifications for the PC's interfaces, inviting other companies to develop peripherals, hardware upgrades, and software. That involved many parties doing many different things to change the technology.

In the last chapter, we saw how dramatically technology changes us, so it seems only fair that we would have influence over it, too. In this chapter we share stories of how people have changed technology by engineering, governing, promoting, managing, investing, and questioning. To unify the stories, we follow the development of nuclear technology.

> In some way or the other,
> each one of us affects the course of history…
> a self-educated Scottish mechanic once made
> a minor adjustment to a steam pump
> and triggered the whole Industrial Revolution.
> – James Burke

Engineer

Most trace nuclear power to the 20th century. We trace it through a self-educated Scottish mechanic in the 18th century all the way back to a Greek in the 1st century to show the engineering process to create ever more useful technology. As a guide, we use a simplified model of a nuclear power plant, separated into three stages:

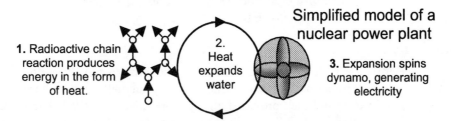

Simplified model of a nuclear power plant

1. Radioactive chain reaction produces energy in the form of heat.

2. Heat expands water

3. Expansion spins dynamo, generating electricity

While step one, the nuclear heat source, is a 20th century invention, step three dates back at least to 1867, when Zénobe Théophile Gramme invented the first practical dynamo. Well before that, Richard Trevithick invented the first high-pressure version of step two, embodied in his 1803 steam railway locomotive. And years earlier, James Watt, the self-educated Scottish mechanic referred to in Burke's quote, patented the steam engine (1769) and actually got it working (1774). Watt refused to use high pressure and, so, his steam engine was a huge, inefficient, stationary machine, useful only in places such as factories.

Watt's steam engine was not the first to perform work with expanding water. He simply improved on Thomas Newcomen's steam engine design, which was on display at the museum in which Watt worked. The adjustment Watt made meant that the engine needed much less coal to do the same amount of work. Still, Newcomen's had been efficient enough to be commercially successful. It pumped water out of flooded mines as early as 1712.

But Newcomen did not invent the steam engine either. He improved on a 1698 steam pump by Thomas Savery, whose barely practical pump also drained flooded mines. And that was a step up from Denis Papin's 1690 concept for moving a piston with steam.

Papin's principle was used in all the steam engines (and nuclear power plants) to follow, even though he technically never built a steam engine.

And Papin may not have known that his concept dated back 16 centuries to a Greek. It was in the 1st century that Hero of Alexandria (also known as Heron) invented a ball that spun by shooting out twin jets of steam. It was no more than a novelty to open doors in a temple, consuming too much wood to be practical for general work.

Still, it must have been impressive to all those kneeling in the temple for the priest to light a fire and have the doors to the altar open magically…and then (with the help of hidden weights) close just as magically when the fire was extinguished. It did not matter that chopping and carting all the wood that it needed was far more work than simply pulling on the ropes to open and close the doors. The miraculous show was well worth it.

At each step, these "inventors" practiced engineering: applying their knowledge of how things work to create something useful. Countless engineers around the world today are making similar advances to create new technology.

> Government is involved in policy direction and
> implementation phases of every major technological system…
> stimulating initiatives, funding research and development,
> regulating for safety and equity, facilitating infrastructure,
> influencing the money market, and functioning
> as a primary customer of high technology itself.
> — Edward Wenk

Govern

Why was Hero's steam turbine developed no farther than a novelty? In 1st century Greece, slaves were an important element of the economy (actually outnumbering freemen by more than two to one) and they provided all the work anyone needed. The slave-based economy would have been rocked by the introduction of technological muscle. Displaced slaves might have caused widespread unrest or even revolution. It may have been an emperor, or an advisor to an emperor, who counseled against developing anything so disruptive as a laborsaving device. Historian Hendrik Van Loon noted, "The amount of mechanical development will always be in inverse ratio to the number of slaves that happen to be at a country's disposal."

And, so, the steam engine played a role in entertainment, but not business. A similar decision was made in Rome at about the same time. The emperor Vespasian is reported to have purchased and destroyed the model of a mechanical device that would have made construction work more efficient, saying, "You must let me feed my poor commons."

In these cases, protecting jobs—and, more generally, preserving stability—motivated government to suppress technology, but government has changed technology in many ways. Rome built infrastructure of roads and aqueducts, directly helping those technologies to succeed and indirectly helping all those technologies affected by dense and connected populations.

Earlier, in the Chapter 3 section on *Protection*, we saw that Venice, England, and the U.S. were among the countries that created patent systems, nurturing innovation by defending intellectual property. Environmental regulations have led to widespread availability of seatbelts and emission controls in cars. But little can compare to the overt role the U.S. took in developing nuclear technology.

[If there had been no Manhattan Project]
Nuclear physicists would have spent years
forming theories and doing experiments
while competing with scientists from other fields
to get money for their work…[but the U.S.]
had plenty of coal and oil…[so] the 1990s
would have had low-power nuclear reactors
operating to produce medical isotopes,
but nothing else.
— Robert Pool

What if governments had decided against splitting the atom and developing nuclear power? Would that technology have remained unknown for centuries, as steam power was? Modern communication is too effective to allow for such a secret, but the costs of making it economically practical might have been prohibitive. It seems unlikely that commercial organizations would have taken on these huge and expensive government projects:

1. **Developing a workable model of a power plant**, giving commercial developers an option for not researching and developing their own designs. In the next section we look at how management decisions influenced the choice of technology for this working model. In the section following that, we look at how industry invested in and further developed it.

2. **Supplying a more-refined fuel** than was generally available. Just 0.7% of naturally occurring uranium is U-235 and 99.3% is the less-reactive U-238 isotope. To refine U-235 from U-238, the U.S. government built a four-story structure covering over 40 acres and a facility with 268 buildings, including laboratories, a distilled water plant and eight electric substations. Developers in countries that have not had access to uranium-235 (U-235) or plutonium have built lower efficiency plants that run on naturally occurring uranium-238 (U-238).

3. **Limiting liability for power plants** to $560 million (the Price-Anderson Act of 1957), no matter how much damage or loss of life was caused. Since insurance companies would not cover more than about $65 million per plant, the government committed to covering the difference. This insurance "rider" of almost $500 million was provided to the commercial operators without charge.

4. **Agreeing to dispose of all the radioactive waste** produced by commercial plants. Since no disposal technology can be proven safe for the tens of thousands of years that the waste will remain dangerous, private corporations might have considered this responsibility too risky to accept. So the government has spent 25 years and $4 billion evaluating and developing the Yucca Mountain storage facility in Nevada. If commer-

cial plants were to pay for it, encasing spent fuel in dry casks of steel and concrete for safe transport to this storage could add $0.03 – 0.06 per kilowatt-hour (kWh) to the cost of electricity. This is a significant sum, considering that their average revenues in 2002 were just $0.07/kWh.

Government gave nuclear power technology a kick-start by figuring out how it could work and by developing the infrastructure to produce the most potent fuel. Then, it kept the technology profitable by limiting risk and promising to take away the nuclear waste. What if the U.S. government had not taken these actions? At the least, nuclear power could have been based on U-238, as Canadian nuclear power is. At most, nuclear power might not have come to the U.S. at all.

In June 2003, the U.S. Senate passed a bill providing increased nuclear research, $865 million to develop new radioactive waste processing techniques, $1.1 billion for a hydro-gen-producing nuclear reactor, and loan guarantees covering half the cost of a group of new reactors. Since then, more bills may have passed.

In democratic countries, the voting citizens influence government action by voting on ballot propositions, contacting their representatives, or making enough of a media splash that government representatives take notice. When that government action affects technology, the responsible citizens are changing technology.

How we choose to influence legislation of such complex issues as human cloning depends on a thorough understanding of the costs and benefits as well as the values we use to measure them. In Chapter 8 we identify some of the common tradeoffs between costs and benefits. In Chapter 9 we make explicit the personal values many use in evaluating technology. Later in this chapter, we will see how a concerned citizen changed a nuclear plant being built near her home by working through government to question its safety.

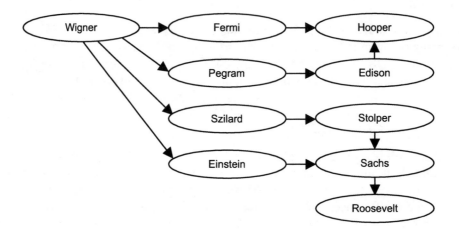

Promote

How does an expensive and complex technology such as nuclear technology actually get started? Not the way a better mousetrap does. An inexpensive and simple technology, such as a mousetrap, can be developed by a lone inventor, who then demonstrates its merit. In the case of nuclear technology, the government involvement necessary to demonstrate it came only after strong promotion.

In 1939, a group of physicists gathered to discuss the possibility of tapping the energy in the nucleus of the atom. Experiments in Germany and the U.S. suggested it would be scientifically possible to release it, which could lead to the technological possibility of a bomb thousands of times more powerful than any in existence. Eugene Wigner, who would later win the Nobel Prize for explaining the structure of the atomic nucleus, believed this so important that the U.S. government must be informed. With government assistance, the technology could be proven impractical or could be developed to counter what Nazi Germany might develop.

The thought of Hitler controlling the only super weapon on earth frightened them all. Wigner convinced Enrico Fermi, George Pegram, and Leo Szilard. Fermi had already won a Nobel Prize for work that would lead to the first nuclear chain reaction, a project he would share with Szilard. Pegram was dean of Columbia University and also a physicist.

How does promotion work? It can work very much like a nuclear chain reaction, with one human contact triggering another. As diagrammed at the beginning of this section, Pegram knew the Undersecretary of the Navy, Charles Edison. The group agreed to send Nobel-laureate Fermi to meet with Edison and make the case for government involvement. Unfortunately, Edison was unavailable, so Fermi met with Admiral Stanford C. Hooper, who spoke a language different from that of the scientists. Hooper wanted to know if he should divert resources from one part of the war effort to a more promising front. Pegram's letter of introduction wanted to cautiously describe the possibilities:

> Experiments...reveal that conditions may be found under which... uranium might be used as an explosive that would liberate a million times as much energy per pound as any known explosive. My own feeling is that the possibilities are against this, but my colleagues and I think that the bare possibility should not be disregarded.

Fermi and Hooper met, and the meeting went nowhere. But in nuclear chain reactions, some high-energy particles go flying off into space without effect. To sustain a reaction, there must be enough other particles that do hit their target. Regrouping, the scientists prepared a new particle, contacting Albert Einstein, whose reputation then was even greater than it is now.

A problem with the first approach to promotion was that nuclear technology was more strategic than tactical. Years of work would precede anything practical, but then the results could change the course of the war and, perhaps, the course of all wars. Government administrators are tasked with focusing on the tactical, the practical steps that can lead to predictable results. This proposal was so grand, the group decided to go straight to the top. Einstein agreed to write a letter to the president of the U.S., Franklin Delano Roosevelt.

But how would they get to the president? More chain reactions. Szilard knew Gustav Stolper, an economist who knew Alexander Sachs, a sociologist, economist, and speech writer to the president. Sachs listened to the scientists' argument and agreed that the possibility of a doomsday weapon was so serious that the president must be informed.

Sachs was a crucial link in the chain. He met with Roosevelt personally and told the story of an American inventor who proposed to French Emperor Napoleon Bonaparte a new weapon invulnerable to weather: ships without sails. Napoleon thought that an impractical dream, but this did not stop the American, Robert Fulton, the eventual inventor of the first practical steamboat. Within 10 days, the U.S. government held the first meeting of the "Advisory Committee on Uranium."

The birth of the Manhattan Project, which created the first practical nuclear technology (atomic bombs), would require many more chain reactions of people connecting to people and selling ideas. Throughout the process, there are key roles:

- Those who develop new information (e.g. physicists conducting experiments)
- Those who know other, important people
- Those who sell ideas, as Sachs did to Roosevelt

Promotion relies on all three. Once an idea gains momentum, however, there is need for a new role: the manager.

Even the Typewriter Needed Promotion

The typewriter, one of the biggest advances in written communication since the printing press, had a rocky start. One might think that the typewriter's advantages over the four-century-old printing press and even more ancient handwriting would be so obvious that no promotion would be necessary. The printing press took hours to set up, with a skilled type-setter carefully placing lead blocks with symbols in relief, so it was practical only for making many copies. The typewriter, by contrast, could uniformly print something even faster than someone could handwrite it.

After the U.S. Civil War, Christopher Sholes invented the typewriter and convinced the Remington firearms manufacturer, which was looking for something to sell in peacetime, to make it. By 1873 they were and, in 1874, Mark Twain bought one to write *Tom Sawyer*, the first typewritten manuscript submitted to a publisher. With a clever inventor, powerful manufacturer, and prominent early adopter, what could possibly stop its success?

The consumer. Between 1874 and 1880 Remington could sell only 5,000. The initial market had been the home, where it failed because handwritten letters were considered personal and civilized. Typewritten letters looked about as personal as a "garage sale" flyer posted on a telephone pole. Some considered typewritten letters condescending, implying that the recipient could not read handwriting.

So, Remington promoted the typewriter to a different consumer, the business market, where efficiency, legibility, and accuracy were priorities. Sales reached 50,000 over the next six years.

Although the typewriter changed little, continued promotion eventually penetrated the home market, too. Popular opinion turned to accept typewritten correspondence to be as personal and respectful as handwritten. First the business and then the home consumer made the typewriter a success. Later, these two markets did the same for its successor: computerized word-processing.

Invention is a flower, innovation is a weed.
— Bob Metcalfe

Manage

The inventor nurtures a technology like a gardener cares for a flower. Inventions that survive to become practical innovations become subject to harsher treatment. Managers' choices favor the tenacious, well-adapted weed over the elegant, but delicate, flower. There is no fate ordained for a technology during the invention phase. Managers can send it in any of countless different directions.

By 1946, several nuclear power technologies had been invented, but none was a working product. Graphite-moderated, liquid-metal, light water, and gas-cooled reactors all had potential. Which ones would thrive? The person who would influence the outcome was a manager, not an inventor. Though he did not even claim to favor the best technology, his choice became the choice of nearly every commercial nuclear plant in the U.S.

Captain Hyman Rickover (later to become admiral) headed the U.S. Navy team that developed a nuclear submarine. The Navy was keenly aware of the problems of powering a submarine with diesel and electric batteries (which we described in the chapter on how technology works). Diesel requires lots of oxygen and produces exhaust—impractical for stealthy, submerged operation. Electric batteries provide very limited energy, quickly needing to be recharged by the diesel. By contrast, nuclear fission requires no oxygen, produces no exhaust, and its radioactive source stores energy a million times more densely than fossil fuels (the last being an advantage for the navy's surface ships, too).

The question, then, was which nuclear technology to use. Rickover weighed his options. He discarded the graphite-moderated reactor because it required more room than the others. Size is less of a concern for nuclear plants on land, but everything on a submarine must squeeze into as little space as possible. The gas-cooled reactor never made it from plans to prototype.

Initially, Rickover also decided against liquid metal (e.g. sodium) because, although it can be more efficient than light water, it is more dangerous—contact with water causes sodium to combust. The complexity of safety systems for liquid metal made light water the less risky choice for the world's first nuclear-powered submarine, the Nautilus, which launched in January of 1955

Still, liquid metal was worth a try for the second, the Seawolf, which launched six months later. But after several crippling leaks in the Seawolf's system, Rickover had its liquid metal reactor replaced with a light water one. Rickover and the Navy had made their choice for a technology that they never claimed was the best, but was simply the most practical and reliable given their specific needs.

That effect on technology spread when President Eisenhower cancelled plans for a nuclear powered aircraft carrier and Rickover was assigned to develop the first commercial nuclear plant. Because the light water reactor had worked so well for him on submarines, he specified a larger version for Shippingport, Pennsylvania. During its construction, he shared information through technical reports and seminars. So, by 1957, when Shippingport started full operation, far more was known about how to build a light water reactor—and what mistakes to avoid—than about building any other nuclear power technology. When General Electric (GE) and Westinghouse invested in commercial nuclear power, they chose to develop the well-documented light water technology rather than the untested and risky alternatives. Hyman Rickover, through his management choices, changed technology.

Civilization is moved forward by restless people,
not by those who are satisfied by things as they are.
— Ralph Cordiner

Invest

Ralph Cordiner, chairman of GE, was restless. In competition with Westinghouse to produce nuclear power plants, GE recognized in the early 1960s that getting big fast was important. The more plants sold, the more experience they gained in building plants, which would lower construction costs. Also, fixed costs of research and testing could be spread across more plants. With lower costs per plant, they could even undercut competition with lower prices while increasing profits.

In this race, would GE or Westinghouse be the one to get big first? Or would a nuclear power technology completely different from the light water reactors that GE and Westinghouse built steal the show? With Europe experimenting with technologies potentially superior to light water, the pressure was on.

To figure out the cost of constructing nuclear plants, GE extrapolated from its experience building fossil-fuel power plants. First, the more fossil-fuel plants it built, the better it got at building them. And, second, the larger the plant, the more economical it was because doubling its output had much less than a doubling effect on the costs of real estate, operating crew, instrumentation and control systems.

GE made its bid to dominate the market by offering construction of nuclear plants at a fixed price (rather than force its customers to bear the uncertainty and risk of this new technology). And it won a $68 million dollar contract to build a plant at Oyster Creek, New Jersey. GE figured it would learn so much that, even if it lost money building this plant, it would make up the profit on subsequent contracts. At 515 megawatts, Oyster Creek was three times larger than any nuclear power plant yet built, and GE offered fixed prices for plants up to 1000 megawatts.

Westinghouse knew that committing to set prices for complex, never-before-built technology like large nuclear power plants involved a lot of risk. However, it, too, understood that the first company to perfect construction techniques could control the market. With hopes of future profit, but expecting the first plants to be loss leaders, Westinghouse jumped in with its own fixed price list. Whichever company blinked first could have only their loss leaders as reward. The other, presumably, would control the market and reap compensatory profits far into the future.

Between 1963 and 1966, the two companies battled for market share. It lasted no longer. They both blinked because the plants were costing them twice as much to build as they were charging their customers. The complexity of nuclear power plants and the potentially far greater catastrophe that could come of failure (and, therefore, the far more restrictive the safety regulations) made them much more expensive to build than fossil-fuel power plants. Both companies returned to supplying just the reactors (a technology with which they had far more experience) at a fixed price, or building complete plants, but on a cost-plus basis.

GE's and Westinghouse's investments brought countless advances in the physical components of nuclear power and in the techniques of constructing them into a plant. These created what we now consider "mature" nuclear power technology. But mature technologies can bring complacency and, as we will see, it can take an outsider to bring important changes.

> The utility seemed reluctant for people
> to look at what they were doing.
> — Juanita Ellis

Question

Before she read about nuclear power in a gardening magazine, Juanita Ellis knew almost nothing about it or the Comanche Peak nuclear power plant being built near her home in Dallas, Texas. Scientists and engineers designed the plant, so what would someone with only two quarters of college education have to say about how it should be changed? Quite a lot, as it turned out.

The article in the gardening magazine did not turn Ellis into a nuclear physicist or engineer, but it did inspire her to question its safety. First, she contacted Bob Pomeroy, the author of the article, to find out more about his concerns regarding Comanche Peak. She shared them, concluding that Texas Utilities (TU) Electric was hiding something about their project. So she formed the Citizens Association for Sound Energy (CASE) with Pomeroy and four friends, attended public hearings, and voiced her concerns to the local media.

Ammunition for CASE's activism arrived with Mark Walsh in 1982, almost a decade after the issue of the gardening magazine that started her cause. Engineers make sure nuclear plants are constructed according to a host of safety guidelines, and Walsh had been one of those engineers at Comanche Peak until his bosses ignored his warnings about defective pipe supports. If a support breaks then a pipe can break and coolant may not reach the reactor core. If not caught quickly, the core can overheat and even meltdown. That is why Walsh quit and sought out Ellis. Other whistleblowers followed and Ellis, representing CASE, caught TU Electric by surprise at subsequent public hearings.

"The little old housewife," as some TU Electric officials called Ellis, was not supposed to understand technology. But her role was not to out-engineer the engineers. It was to focus the technical knowledge of others, like Walsh, in a public and forceful way. The 445-page report on pipe supports that CASE delivered to the hearings caused a lengthy review of construction. Eventually, TU Electric:

- Made the recommended fixes and improvements
- Compensated CASE for its research ($4.5 million)
- Compensated 50 whistleblowers fired from plant ($5.5 million)
- Invited a CASE representative to the plant's independent safety review committee

The fight was not painless. Ellis invested a lot of time. Pomeroy was labeled a subversive in a profile developed by the Texas Department of Public Safety. Comanche Peak took almost a decade longer than planned, due to the wrangling in the public hearing. Also, the plant cost $11 billion, substantially more than the $779 million originally estimated. But without Ellis, Pomeroy, and the battle they waged, it is possible Comanche Peak could have earned the sort of fame that Three Mile Island or Chernobyl received (we analyze the Three Mile Island accident in the next chapter on the costs and benefits of technology). Through questioning, they changed the process around the physical technology.

How the approach in this book changes technology

Consuming, inventing, governing, managing, investing, and questioning are just a few of the ways we shape technology. We can also change it by teaching about these ways. That multiplying effect is what this book aims for and what Don Jacobs has accomplished in his classroom.

Don Jacobs challenges his middle school students to lift his 200+ pound body off the ground for 10 seconds. All fail until he retrieves a thick wooden board from a hiding spot and places it across a rounded block. With Jacobs standing on the short end of this lever, even the smallest student can step on the long end to lift him from the ground and keep him there. His students experience a simple and very old technology that extends our abilities.

I met Don Jacobs in Santa Cruz, California, through Ben, who had been a student in his classroom and, later, mine. My first afternoon in Jacobs' classroom, I described a curriculum (developed by KnowledgeContext and available on that educational nonprofit corporation's website: www.KnowledgeContext.org) that parallels this book. He assumed I was talking about computer training—what else could "technology education" be? With two decades of teaching experience, Jacobs recognized the importance of teaching within a big picture, the importance of helping young minds make connections between concepts. And as a crusader for technology, he also was keenly aware of the problems that new technology both encounters and causes.

Working in a self-contained classroom, Jacobs was free to show connections between technology and every subject his students studied, so he tried the curriculum on his class. Then, when he came to a KnowledgeContext board meeting to share his experiences, he carried the authority of someone from the front lines. The message he brought was that the connections of technology to every subject and to the future success of his students was so important, that he believed the curriculum should be mandatory for all students.

Not surprisingly, the computer-savvy students in Jacobs' class were immediately receptive to a big picture for the technology that was already a hobby for them. The curriculum showed them where computers came from, aspects of how they worked, many technologies beyond computers, the broad reasons we use technology, and how to evaluate the costs and benefits.

The curriculum works in my classroom!
It crosses the Digital Divide to get all my students
thinking about the nature of technology.
By the time we finish, they've got a tool
for understanding and evaluating technology.
— Don Jacobs

But the students on the other side of the Digital Divide—those less fluent with computers—found even greater value. The alienation from technology that they felt lifted when they realized that they already use technology throughout their lives. In a lesson that mirrors this chapter, they discovered that computer programmers and "scientists in white lab coats" are just two of many roles that influence technology. The curriculum brought them a broader perspective. They went from being "not as good" as the computer whiz kids to having their own, personal, independent relationship with technology—one that tied into their own personal interests in painting, music, cars, or almost anything. Finally, they had context for technology.

The importance of context is well accepted in pedagogy, and schools build context for many subjects. History provides context for economic changes and development of art. Science provides context for applications of math, and is understood in terms of major themes like energy, evolution, patterns of change, scale and structure, stability, and systems & interaction. The meaning or significance of something emerges from the big picture in which it resides. Critical thinking—understanding and evaluating—relies on context. And context is where teachers are strongest.

> How can I teach these kids
> anything about technology
> when they know so much already
> and have so much time
> to play with it?
> – 7th-8th grade teacher
> (in Silicon Valley)

Teachers are busy planning curriculum, grading papers, handling behavior problems, meeting with parents, and keeping abreast of the educational field. Most are better equipped to present the enduring context of technology than they are to teach ephemeral technical information, such as the installation procedure for a new computer operating system. The ICE-9 curriculum leverages the rich and broad education that many teachers bring to the classroom. It helps to put teachers in the position of providing students with the big, contextual picture, and makes more acceptable that some students may be the "experts" on some technical issues.

For teachers and students alike, ICE-9 can be like a Swiss army knife, nine questions that find utility in a variety of situations. Jacobs pointed out that, while classroom projects for evaluating things like pollution and energy sources are popular, in many classrooms the process boils down to an emotional response aligned with the teacher's personal viewpoints. By contrast, this curriculum gave students an objective strategy to analyze technology-influenced issues and form their own critical positions.

Jacobs' impact ripples out through hundreds of students. A century from now, the biography of someone who dramatically changed technology may trace back to Don Jacobs' inspiration and teaching.

Our deepest fear is not that we are inadequate.
Our deepest fear is that we are powerful beyond measure.
It is our light, not our darkness that most frightens us…
Your playing small does not serve the world…
as we let our own light shine, we unconsciously give
other people permission to do the same.
As we are liberated from our own fear,
our presence automatically liberates others.
— Marianne Williamson

Marianne Williamson's poetry about our power as individuals echoes recent scientific discoveries. Attempts to predict avalanches and earthquakes have revealed that natural systems are often on the edge between stability and chaos. To see how this works, imagine throwing a snowball on a meadow of snow. It lands, affecting a small area in a small way. Now, imagine throwing a snowball at an avalanche. It is consumed, having little impact. Finally, imagine a slope of snow so steep that any additional snow would cause an avalanche. One snowball could have a huge impact.

Intuitively we know that most snow is not on the razor's edge of a devastating avalanche. Otherwise, skiing and snowshoeing would be extremely dangerous sports. But if many tiny mounds of snow were on this edge between the stability of the meadow and the chaos of an avalanche, we might not even notice.

Scientists have found this edge in many less-conspicuous natural systems, using "chaos theory" or "complexity theory" to analyze them. It underlies the story about the beating of a butterfly's wings in one part of the world causing a storm in another. Although the global weather system is very unlikely to be on such a fine threshold that a butterfly could have such an impact, it does illustrate how many smaller changes actually occur.

And now some are applying this scientific approach to a wider variety of phenomenon. In the book *Ubiquity*, Mark Buchanan analyzes mass extinctions, the start of wars, and the fall of empires, finding their historical frequency and magnitude to agree with this model. Civilization treads between meadow and avalanche, with small actions sometimes causing large results.

Were civilization all meadow or all avalanche, no individual could have an impact. And yet the individual does, as was the case in 1914, when Archduke Ferdinand of Austria accidentally turned down the wrong street when visiting Serbia and was assassinated by an armed and angry student who could not believe his luck. So started the First World War.

Unlike that student, few of us tread on the edge of a world war, but in our families, neighborhoods, schools, places of worship, businesses, and government, we encounter many such sensitive edges. Lacking omniscience, we rarely know how close we are to triggering great change…or how things would turn out if we do not take action. All we know for sure is how things do turn out.

Guiding our actions is our model for how change happens. If we believe our world a meadow or avalanche, then we might as well just get by. Extra effort would be pointless.

However, if we believe our world follows myriad edges we can influence one way or the other, we consider ourselves truly powerful.

> For the want of a nail, the shoe was lost;
> For the want of the shoe, the horse was lost;
> For the want of the horse, the rider was lost;
> For the want of a rider, the battle was lost;
> For the want of the battle, the kingdom was lost;
> All for the want of a nail.
> — Benjamin Franklin

How do we change technology? By nudging it off edges in the direction of our choosing. With billions of tiny nudges, consuming some technologies and disdaining others. With brilliant inventions. By governing, legislating, adjudicating, and voting. Through management choices maximizing utility and minimizing risk. By investing time and capital. With questioning of assumptions. By teaching about these roles and the greater context of technology and how we evaluate it.

We change technology in more ways than this, and all are guided by our understanding of it. We are its creator, beneficiary, victim, judge, and guide. With a contextual understanding of how it connects with all aspects of our lives, we can make conscious, deliberate choices. We can create an intentional future.

Chapter 8

What are Technology's Costs and Benefits?

It is usually difficult or impossible
to manage the effects of an innovation
so as to separate the desirable
from the undesirable consequences.
– Everett M. Rogers

Technology has both costs and benefits. Every technology. The production and control of fire has protected us from predators and from the cold. But it has also burned our fingers and our homes. Clearly, costs and benefits are measured in many ways other than money: time, jobs, fairness, sustainability, environment, advances in knowledge, culture, and more.

For instance, while motor vehicles cost money to acquire and operate, enabling us to do things that generate money, they also cost about 40,000 deaths per year in the U.S. Air pollution and dumps full of tires and crushed cars are environmental costs, but cars provide environmental benefits, too. Prior to the car, city streets were filled with working horses, each producing an average of 22 pounds of dung per day. In 1900, the 150,000 horses in New York City and the 15,000 in Rochester produced over half a million tons of manure—enough to form a block 200 feet on a side and 190 feet tall. At the time, the horseless carriage appeared quite an environmental improvement.

Often it seems that the more we get of one desirable quality, the more we lose of another. These tradeoffs transcend the details, and new technology simply allows us closer to the extremes. For example, consider how easy it is to become dependent on something very useful. And the more useful a technology is, the more we rely on it—just like a crutch. But unlike with a crutch, we don't plan to "get better." The Y2K threat reminded us how much we lean on computers to process our financial transactions and run our businesses, and we do so precisely because they are so useful. How much will we rely on future technology far more compelling and enabling than anything we know today?

An unpredictable crutch is more dangerous than a predictable one. Unfortunately, many of our enabling technologies are complex in ways that make them hard to predict. When it comes to generating electricity from radioactive material or to shooting down enemy missiles, predictable is very important. In this chapter we will see why these systems, in particular, have already surprised us with failures—and why we continue to make ever more complex technologies despite the risks.

Such failures can be sudden and dramatic or subtle and lingering. "Catastrophic" and "chronic" describe costs and benefits in general. We often trade one for the other. Fires are a catastrophic event. Inhaling the asbestos that prevents our homes from burning causes chronic problems. Heating our homes or generating electricity with coal pollutes the air and can also cause chronic problems. Doing so with nuclear power keeps the air clean, but can cause catastrophic problems if a meltdown of the reactor core occurs.

Control and freedom is another pair of linked costs and benefits. After the terrorist attack on September 11, 2001, many Americans were ready to trade freedom and privacy for control and security. Technology, of course, was ready to accommodate. More than 500 years earlier, Gutenberg's printing press spread freedom of speech while undermining the control of the Catholic Church. Some technology can be used either way, but using it to increase either control or freedom often decreases the other.

Finally, it seems inevitable that the faster our technology progresses, the faster it renders itself obsolete. Cellular telephones shrink while gaining new features. Most don't live to see their 2nd birthdays because newer models lure their owners away. The cost here is environmental, with the creation of thousands of tons of garbage each year, but other forms of technological progress can cost jobs and culture.

These patterns do not alleviate us of diving into the particulars of issues that concern us. "The devil is in the details," and there is no substitute for that analysis. But how do you put those details into context? How do you decide which details of the uncounted mountains of them to explore? These general patterns describe the big picture. They provide that context and can serve as a map. Future technology may make our choices more complex and difficult, but these patterns will weave through their details, too.

> So, you might say let's just back off from science and technology.
> Let's admit that these tools are simply too hot to handle…
> throttle back to a minimal, agriculturally intensive technology,
> with stringent controls on new technology…
> or you might imagine throttling back
> much further to hunter-gatherer society.
> – Carl Sagan

Enabler vs. Crutch

Hunting and gathering gave way to agriculture in most parts of the world. Each generation of farmer remembered less of the foraging ways. Population grew so dense that there was no going back. Even if all the skills and knowledge of hunting and gathering could be remembered, it could not support ten to one hundred times as many people in the same area. Agriculture enabled us to multiply and specialize, but not go back.

Writing enabled us to remember far more outside our brains than inside. And, like a crutch, it allowed our memories to lose their edge. Computers have gone farther, making it ever easier to look up facts when we need them and no longer memorize them. By linking computers, the Internet has made it easy to search the world over for more facts than anyone could ever retain. The more compelling a technology is, the more likely we will grow dependent upon it.

Dependence is not necessarily bad. We can apply our brainpower to understanding relationships, connections, and processes—including the processes for looking up those aforementioned facts. Well, perhaps this does not exonerate dependence, but simply asserts the obvious: that the benefits can outweigh the costs. Less obvious is that, even when they do, the costs are still lurking, and can still be dangerous.

For instance, it seems an appropriate use of computers to free countless humans from simple calculations. After all, computers can add millions of numbers in a second without a mistake and the human brain, so prone to simple errors, is capable of creativity beyond any computer (any computer of the present day, at least). A clear case of the benefits outweighing the costs. Yet, at the end of the 20th century, the Y2K scare forced us to recognize the cost of this otherwise reasonable choice.

In Switzerland, 105-year-old man was directed to attend elementary school when a computer program miscalculated his age to be five. In Norway, 16 airport express trains and 13 high-speed, long-distance Signatur trains refused to start because the on-board computers did not recognize the date—and that happened December 31, 2000, a year after problems were expected. Y2K helped us realize how insidious computers had become. We become reliant on them precisely because they are so effective at certain tasks—whether processing thousands of credit card transactions per second or monitoring our car's wheels for skidding 45 times each second.

The Simple Reasons Behind the Y2K Bug

The Y2K problem arose from a memory-saving shortcut of storing only the last two digits of calendar years in our computer programs. Why repeat "19" a billion times in valuable computer storage if, when printing or displaying, automated programs could simply prefix it to the two digits that we do store? In the 1970s and 80s (that's **1980s**, of course), it seemed a safe assumption that software would be replaced well before 2000. Besides, profits depended on saving money "this year," not decades in the future.

We discovered that not only did old "legacy" software survive until 2000, but new software often reused standard subroutines containing the two-digit shortcut. These basic building blocks of software perform tasks common to many programs, such as manipulating or comparing dates. Like pouring a bit of old milk into a fresh carton, reusing old subroutines can make the whole program go bad.

The problem arises when computers perform arithmetic on dates, figuring out if your car's engine is overdue for service or your credit card or mortgage have been paid on time. January 2000 minus January 1999 is one year, but January 1900 (which is what a program that stores only the last two digits would assume) minus January 1999 is not. Is it –99 years? That depends on how the computer software is written, and many of the original programmers who could answer that question were long gone. In many cases, we simply did not know what would result.

We escaped the worst predictions for the Y2K crisis, spending millions of dollars on preventive measures, such as consulting and system upgrades, and experiencing only scattered system failures. But, though we averted disaster, we were forced to recognize how dependent our society had become on technology. On November 9, 1965, when an electric power relay at Niagara Falls, New York, switched off, the reminder was not as gentle.

That switch triggered a series of events that, within minutes, plunged 30,000,000 people over 80,000 square miles into darkness. The switch operated properly, protecting a power line from overload, but its threshold was based on two-year-old power levels. In those years, consumption had risen to the point that a momentary surge in demand triggered the mechanism. Taking one power line out of service, burdened the remaining lines with an extra load, triggering their power relays. Like dominos, the parts of the power system knocked each other offline.

Deprived of electricity, subway trains stopped and sat in pitch-black tunnels. Traffic lights stood dark over impossible traffic snarls, the orchestrated movements of cars and trucks thrown into chaos. Half of the 150 hospitals were not equipped with emergency generators, so they went dark. Highly flammable gases used in surgery prohibited use of candles or any open flames to finish what surgeons had started.

Airports, similarly unprepared, disappeared into the night. Approaching planes searched for the landing lights and listened for the control tower, unsuccessfully. Some basic communication was reestablished from planes parked on the tarmac, as their radios were powered by on board batteries.

Growing more and more specialized over the millennia, we have moved from self-sufficient hunters and gatherers to cogs in a "great machine." If the machine stops, we wait until it starts again because, with few exceptions, we are unprepared to operate outside of it. This disaster illustrates how dependent we are on a complex structure of technology so tied into our lives that we are hardly aware of it.

> Where once we lived isolated and secure,
> leading our own limited lives, whose forms
> were shaped and controlled by elements
> with which we were intimately acquainted,
> we are now vulnerable to change which
> is beyond our own experience and control.
> Thanks to technology no man is an island.
> — James Burke

Back to the fateful day in 1965. 800,000 subway passengers waited for rescue. Those stuck in traffic waited. Those at home, waited, hoping that the food in their now-dormant refrigerators would not spoil before everything started working again. They probably did not consider that the food in their cupboards could run out before local stores could be restocked by trucks that bring food from ships and trains—all part of that great machine. The 250 airplanes planning to land at John F. Kennedy Airport diverted to airports outside the affected area. They could not wait.

Power was back by early the next day, leaving a few dead and many shaken. The tens of millions who experienced the blackout had received a harsh wakeup call. Any illusion that we could get along without our technology was gone.

Was it a mistake to allow ourselves to become so reliant? If so, where would we draw the line? No computers? No electricity? No writing, agriculture, or any technology at all? Even our spear-wielding Cro-Magnon ancestors would balk at that.

Avoiding reliance on technology is rather hard—and unpleasant—to imagine. It allows us to accomplish so many things so much more easily. We make use of it and stop doing things the old, inefficient way. Those of us that "did it the old way" age and die. Institutions that supported the old way crumble and disappear. Our bridges are burned.

The solution is not to go all the way back to a time before technology, but to recognize what choices we are making. In the 21st century we have more technology choices than ever. Whether those choices are individual, organizational, national, or global, we can consciously weigh the costs and benefits. What are the dangers of becoming dependent on a given technology? Does it enable us in ways important enough to justify that dependence? If so, can we mitigate the risk of dependency? And does mitigation involve other technology with similar issues?

In retrospect, could we have prevented the failure of the power relay? Could we have avoided the shortsighted decision to store only the last two digits of each year in computer files? Knowing what we do now, of course we could have averted these specific problems, but without the benefit of hindsight can we avert future failures? The answer becomes more and more important as ever more enabling technology fosters ever greater dependence. Unfortunately, the systems of our great machine continue to grow ever more complex…and that can make them very difficult to predict.

> Things should be as simple as possible,
> but not any simpler.
> – Albert Einstein

Complexity vs. Predictability

There are two reasons that complex systems can be hard to predict. The first is a rather obvious one: complex systems are harder to understand than simple ones. And, if they also happen to have more points of potential failure, all the worse. Clearly, we cannot predict what we do not understand unless we can observe its full range of behavior. For instance, the sun rises and sets each day. Simple. We need no scientific understanding of the sun to predict that, but we cannot do the same with a nuclear plant. Building one just to see if it blows up is impractical.

The second reason also has to do with behavior and is based on a concept that came up in our discussion of how technology works in Chapter Four: emergent behavior. With the right ingredients, a system can behave quite differently from any of its components. If it is so complex that an expert can understand only one component, then bringing together experts, each versed in one component, to predict the behavior of the whole would not be very helpful.

We could avoid creating complex systems altogether. That would free us from this source of unpredictability. But the benefits of a complex system can be alluring and may free us of other costs. Take air travel, for instance, with its ability to transport us thousands of miles in just hours.

Alternately, we could drive a car, but at what cost? Death rates per mile traveled are 100 times higher for driving than flying. So, in spite of the complexity of airplanes and air traffic control, we continue to choose it. For as we explored earlier, returning to simpler technology means giving up a lot.

With flying so much safer than driving, maybe this tradeoff between complexity and predictability is more theoretical than practical. The key question: do our complex systems really fail when we do not expect them to? If so, do these failures affect us?

In 1974, the U.S. government released the "Rasmussen Report" to document the safety of nuclear power. Under the direction of MIT nuclear engineering professor Norman Rasmussen, the report was based on careful, systematic analysis of two reactor sites that represented U.S. reactors. Rasmussen and his team were familiar with emergent behavior, and they considered how several minor problems could combine to cause a major one. Their scientific approach estimated probabilities for each type of occurrence and for a variety of their combinations.

The conclusion: a meltdown could occur once in every 17,000 years for each reactor, or roughly every two centuries, based on the number of reactors in the U.S. This meant that based on the 107 reactors the U.S. had by 1998, a meltdown would be predicted every 159 years. Less than five years later, it happened.

Middletown, Pennsylvania: Three Mile Island nuclear plant. At 4:00 AM on March 28, 1979 in Unit 2 of the plant, a combination of malfunctions and mistakes shut down the pumps circulating water through the steam generator (loop B in the diagram of a reactor). The heat in the water that flows through the uranium reaction (loop A) had no place to go, so the reactor core heated up. Heat sensors detected the increasing heat and water pressure, automatically shutting down the reaction by inserting control rods into the core. Also automatic was the opening of a relief valve, which removed the high-pressure steam (from loop A).

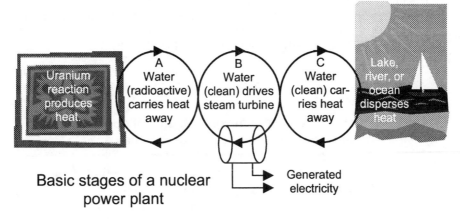

Basic stages of a nuclear power plant

Both automatic responses were correct, but the relief valve failed to close once the steam had been bled off. Now, instead of excess pressure, the water circulating through the core had inadequate pressure. Two high-pressure injection pumps came on automatically, which led to the next problem. Thinking that they were doing the right thing, plant operators turned off one pump and turned down the other. Put simply, the operators had misunderstood what some of the 600 warning lights were telling them. Sometimes too much information can cause as much trouble as too little.

In this case it caused more. At approximately 5:30 AM, operators turned off the pumps circulating water through the reactor core (loop A), still believing that they were following the proper procedures. Deprived of water, half the reactor core became exposed, melting and releasing radioactive material. Around 6:30 AM, they figured out that the valve had not closed (loop A), which had been the second problem. They restored water to the reactor and started the process of getting it back to normal temperature and pressure.

The Rasmussen Report had not anticipated that particular interaction of independent events or the short time it would take to encounter them—less than five years after the report was issued. While the operators could be blamed for much of the problem, the complexity of the system was also responsible.

With systems monitoring systems, there are many modes of failure and too much information for the operators to absorb (including the 600 warning lights). The safety systems that monitor other safety systems add complexity, offering new ways that failures can occur.

The real problems we encounter tend to be more complex than the theoretical problems we anticipate. Fact is stranger than fiction. In *Beyond Engineering*, Robert Pool shares the observations of Hyman Rickover, someone "who oversaw the construction of more than a hundred reactors." Rickover contrasted nuclear reactors in theory ("on paper") and in practice:

On paper	In practice
It is simple	It is complicated
It is small	It is large
It is lightweight	It is heavy
It is in the study phase	It is being built now and is behind schedule
Little development required	It requires an immense amount of development
(Uses off-the-shelf components)	*(Much of it on apparently trivial items)*
It can be built very quickly	It takes a long time to build
	(Due to engineering development problems)

> The more we study the major problems of our time,
> the more we come to realize that they
> cannot be understood in isolation.
> They are systemic problems, which means that
> they are interconnected and interdependent.
> – Fritjof Capra

Another example of a system so complex that we failed to predict its behavior became infamous during the 1991 Persian Gulf War when a SCUD missile launched from Iraq crashed through barracks in Saudi Arabia, killing U.S. soldiers. The soldiers were protected by the Patriot missile, a sophisticated technology that had worked very well...right up until it missed that SCUD. While some tasks are so difficult that a number of failures are expected, missing the missile was a tragic surprise.

Why Did the Patriot Fail?

What caused the Patriot missile to miss the Iraqi-launched SCUD missile that killed 28 soldiers? A mathematical error that did not show up when the missile was tested. Why did it show up in combat? The radar tracking system that was supposed to guide the missile to the incoming SCUD had been running for 100 hours, but had been tested and certified for only 14 hours. If it had been shut down every 14 hours, the mathematical error would not have grown large enough to matter.

But why should it matter, whether the radar system ran for 14 hours or 100? To answer that, we need to poke inside the Patriot, which involves radar and voting for ice cream.

Radar works by transmitting an electromagnetic wave and then timing how long it takes to bounce off an object and return. That delay is translated into the direction and distance of the target, so it must be fairly accurate. The Patriot system counted this delay in tenths of seconds, but computers represent information in binary and the binary equivalent of one tenth is not exact. The difference is like the one in representing fractions in terms of percentages.

Ask three people what their favorite ice cream flavor is. Chocolate, vanilla, and strawberry each gets one vote, or 33% of the total. Add all the votes to get 33 + 33 + 33 = 99%. One percent is missing. If each flavor gets 33.3% of the vote, this still adds up to only 99.9%, so one tenth of a percent is missing. Keep adding precision to make the error arbitrarily small, but at some point that process costs you effort and costs the system memory and performance. At some point, the error is just too small to be significant.

The missing delay in the Patriot's radar system was very small, but it accumulated over time. Over 14 hours, it did not affect performance, and since 14 hours was as long as the system was specified to run between shutdowns, it was not significant. After 100 hours, it was 0.34 seconds. That was significant. Traveling at three times the speed of sound, a Patriot missile goes about 1000 feet in 0.34 seconds. And whether the Patriot was 1000 feet beyond it or short of it, the SCUD had plenty of room to get by.

The Patriot is a sophisticated and complicated technology, making it difficult to predict. Because it consists of many parts developed by many people over a long period of time, it can fail in ways that no single person is aware of. The original system was designed to track airplanes. The engineers who redesigned it to track missiles, which are much faster, may not have known all the assumptions that the original engineers made. In a simple technology, this would not have mattered. The redesign would have been straightforward because all the parts of the system—and their interactions—would have been apparent.

But there was no simple alternative. The best missile defense the U.S. had at that time was the Patriot. Given that the U.S. was at war with a country that had surface-to-surface missiles, the Patriot was the best option. Ironically, a software fix to the timing problem had already been developed and distributed, but it arrived after the SCUD struck.

If nothing else competes for your time and money, additional testing is always good. Of course, that is never the case. The complexity and dangers of a technology can suggest how much testing we should do. That testing effort becomes part of the technology's cost...and part of our evaluation. Sometimes, we evaluate the complexity and dangers as so great that we do not pursue a technology, as in the following case of biotechnology.

Baboons do not contract AIDS. In 1995, Dr. Steven Deeks suggested that injecting elements of baboon bone marrow into humans with AIDS might resuscitate their impaired immune systems since those system rely on cells produced in bone marrow. With many lives on the line in the U.S. and even more outside, this was a serious proposal. Delaying an effective treatment meant needless loss of life.

Evaluating this proposal, the medical community recognized a complex, unpredictable system. The rapidly mutating human immunodeficiency virus (HIV), the genetic diversity of patients, and the unknown viruses that might be lying dormant within the bone marrow of baboons formed an interacting biological system that everyone agreed was beyond our ability to predict. Since serious diseases have jumped from animal to human hosts, mutating as they go, this test ran the risk of introducing a new one.

To grasp even more clearly why the risk was so high, let's explore this in a bit more detail. HIV jumps from monkeys to humans. The influenza virus develops in birds, which are not affected, and then spreads to pigs (wallowing near and in bird feces), which allow it to mutate into a form that can jump to humans. That mechanism had evidently changed when several people died in Hong Kong from a flu that jumped directly from chickens to humans (no pig incubator necessary). What do you do when a complex system suddenly changes and you can no longer predict its behavior? Epidemiologists pushed for the annihilation of all chickens in Hong Kong before it could spread farther. After this drastic step, that particular flu strain disappeared.

But HIV and influenza are not the only diseases known to jump from animals to humans. Long ago, once humans domesticated and started living in close proximity to them, cattle probably gave us measles, tuberculosis, and smallpox. With all this in mind, health professionals worried that an HIV-infected human could prove an easy training ground for a baboon-hosted virus. If it existed, it could venture from the baboon's marrow out to the injected human's body, trading pieces of genetic material with other viruses, mutating into forms unknown.

Nobody could predict if the proposal would save lives or if it would create a plague far worse than AIDS. Even though the plague scenario seemed extremely unlikely, the compromise agreed to was to test the baboon marrow on only one patient, so that it could be better controlled. The result was anticlimactic: the baboon's bone marrow did not grow in the patient's body, perhaps rejected by the remnants of the patient's immune system.

So far we have reaped neither the cost nor the benefit of that medical technology. But this area—and biotechnology in general—presents us with far more complex and hard-to-predict systems than we have ever encountered. Living systems do not have emergent behaviors as a side effect, as some of our mechanical technologies do, but as a distinguishing feature.

This does not mean that systems as complicated as nuclear reactors, missile defense, or medical technology are inherently unsafe. It does mean that complex systems are very difficult to predict and, so, we have had to learn by experience. In human evolution, the first compound tools (i.e. created from multiple parts, such as lashing an axe-head to a handle) must have caused some surprises. With experience, we learned to predict those pretty well. With time and, sometimes painful, experience we will learn to predict more complex systems pretty well. Currently, computer simulation show promise in this, but before that technology gets our current complex systems under control, we are already off creating technologies even more complex. This cost/benefit tradeoff of complexity vs. predictability is a pattern that will not go away.

In the majority of cases,
the evil is very insidious...
The worker falls into ill-health
and sinks away out of sight
in no sudden or sensational manner.
— British asbestos textile factory inspector, 1898

Catastrophic vs. Chronic

Fire is a catastrophic problem. In buildings, it suddenly takes many lives each year, and has since humans first brought it into their early abodes. While caves, adobe, and igloos cannot catch fire, many structures are constructed of materials that can. Asbestos, a fibrous mineral, was used extensively as a fireproof insulation because it resists burning. Its resistance to heat also made it an ideal material for brake pads on cars, trucks, and trains. A vehicle that does not stop can also be catastrophic.

Fibers from asbestos can scar lung tissue and cause asbestosis. Asbestos exposure has also been linked to cancers of the lung, larynx, pharynx, oral cavity, pancreas, kidneys, ovaries, and gastrointestinal tract. In 1950 an internal report from the chief physician of Johns-Manville, a major asbestos manufacturer, said that "the fibrosis of this disease is irreversible and permanent so that eventually compensation will be paid to each of these men but as long as the man is not disabled it is felt that he should not be told of his condition so that he can live and work in peace and the company can benefit from his many years experience."

These are chronic conditions that take many years of exposure to develop so insurers of the asbestos industry argued over who should pay the victims: the insurer at the time of the exposure or the insurer at the time of the diagnosis and claim. The death of the asbestos industry in the U.S. was a messy business near the end of the 20th century, with former workers often dying before courts awarded compensation. What caused this chronic problem in the first place? Fixing the catastrophic problem of fires and failed brakes. Without fully understanding the costs and benefits, we traded catastrophic problems for chronic.

...risks that are under someone else's control,
potentially catastrophic and unfamiliar
are perceived as greater than those with
the opposite features. That is why most of us
view riding our bicycle in a busy street as
a more acceptable risk than living near
a nuclear power station, although rational
analysis says that you should stay off the bike.
— John Krebs

The roots of the word "catastrophic" come from Greek, with "cata" meaning down, against, back, and "strophe" meaning a twist, or a turning about. Although it is often used to describe costs, we can take it in its more general meaning of something great and sudden. "Chronic" comes from the Greek "of time." It, too, is commonly used to describe only costs, but we use it for benefits, too.

Trading catastrophic for chronic, or visa versa, is a recurrent theme in the history of technology. Dying of cold in the winter is catastrophic. So burn coal. But the effects of coal pollution (lung cancer, acid rain, and global warming) are chronic. So you decide to split the atom. Chernobyl ran clean until operators disabled the automated safety systems to try out a new emergency procedure. In just five seconds, the reactor's power skyrocketed to 500 times normal. It melted, caught fire, and exploded, spreading radioactive material into the air. That is catastrophic.

We tend to fear catastrophic costs we cannot control more than the chronic ones we can. So, while the effect of many diseases can be catastrophic—death—the result of misuse of antibiotics to fight those diseases can be chronic. Specifically, antibiotic-resistant bacterial strains are a chronic problem, which we create by either overuse of antibiotics or failing to complete the prescribed course (e.g. stopping halfway through a 60 day prescription because we started to feel OK or ran out of pills).

Catastrophic costs are often easier to predict than chronic. With 40,000 deaths from car accidents in the U.S. each year, we have plenty of statistics to project into the future as well as a "smoking gun" cause of those statistics. Asbestos is not as easy to identify as a car when it causes a death because over the years it takes to kill many other factors affect the victim. Medical science can now readily identify the fingerprints of asbestos-related disease, but other chronic diseases are still very hard to attribute and, consequently, to predict.

Modern life exposes us to pesticides and drugs in our foods, as well as pollutants in our air. Could these cause or aggravate health problems such as asthma? According to the CDC, "Asthma is a complex disease that is increasing in prevalence in the United States. Poor, inner-city minorities have disproportionately high rates of mortality from asthma. We still don't know what causes this disease or how to cure it…" We are not suggesting that we know better than the CDC what causes asthma. We are suggesting that any chronic costs of technology can be hard to connect to the offending technology and, so, will be very hard to predict.

With new and novel technologies, we will encounter both catastrophic and chronic costs and benefits. The more technology we employ, the harder it will be to untangle cause and effect, especially in the case of chronic costs—or even benefits. As this complexity increases, we will need to remember the simple pattern of tradeoffs that may lurk just beneath technology's glitzy veneer.

We are in a new cycle
[following the 9/11/2001 terrorism]:
We'll trade our privacy to be
more collectively secure.
— Harold J. Krent

Control vs. Freedom

At the end of the 19th century, a 22-year-old Italian man arrived on British soil bearing a small black box he claimed could be used to send telegraphic messages without telegraph wires. Thinking more brilliantly than he was paid to, a British customs official recognized the threat to security—the very stability of the British Empire—if just anyone could move information unseen by the government. Taking the initiative, he smashed the box and sent Guglielmo Marconi back to Italy.

The technology in question—radio technology—ultimately increased freedom by allowing people to communicate more freely. And it undermined government control. Early in the 21st century, the Taliban in Afghanistan destroyed televisions and satellite dishes because they undermined control of information. These technologies could also increase control at the expense of freedom. In the 15th century, Gutenberg's printing press undermined the control of the Catholic Church by increasing freedom of communication.

Little more than a decade after Marconi's prototype radio was smashed, Britain nationalized (and protected) much of the Marconi Company's facilities located within its borders, claiming that wireless communication was a matter of national security. How else could the British navy communicate with its ships at sea? Today, radio coordinates security forces throughout the world. In Britain televisions monitor streets for crime, where the mere presence of cameras, it is hoped, will discourage criminals. So technology can similarly be used to increase control at the expense of freedom.

> Like most powerful technologies,
> total surveillance will almost certainly
> bring both good and bad things into life.
> —James Wayman

The tradeoff of control (and security) for freedom (and privacy) is a pattern in society. Since it so often involves technology, it is also a cost and benefit of technology. It is up to all of us to weigh these because those developing the technologies may not. A manager of a system being designed to identify people by the spectrum of light reflected from their skin—much as satellites identify minerals or camouflaged vehicles—said, "We develop the technology. The policy and how you implement them is not my province."

While Internet technology allows more freedom of communication, working its way around even China's strict controls, it is privacy that is generating the most controversy. Cameras watched people entering the stadium for the 2001 Super Bowl in Tampa, Florida, but unlike conventional systems monitored by people, this one was monitored by computers. Each face was digitally characterized and compared to a database of known criminals. This was just a test run for implementations to come, so nobody was arrested, though 19 probable matches came up.

A U.S. government mandate to incorporate global positioning systems (GPS) into cellular phones is motivated by the importance of locating calls to 911 emergency centers. Calls from landlines—standard home, business, and pay phones—already provide location information on the computer screens of 911 operators. Even if you can't tell them where you are, they know. This would be useful for cellular phones, too. But, though it has been called a matter of safety, it is also a matter of privacy. The same technology would allow anyone making a cellular call to be located and, over time, tracked. Already, the content of calls can be monitored in the name of national security.

Echelon, a mostly-secret cooperative effort joining U.S., Britain, Canada, Australia, and New Zealand, monitors phone calls. Cellular calls are the easiest because they broadcast into the air, but with cooperation from (or the tapping of) telephone companies, most calls can be monitored. As with the cameras at the 2001 Super Bowl, computers are used to analyze the huge volumes of data. Voice and pattern recognition help to identify those conversations that should be analyzed by humans.

Run by the U.S. National Security Administration (NSA), Echelon has company. The Federal Bureau of Investigation (FBI) developed Carnivore (also known as DCS1000) to monitor email, subject to a legal search warrant. The system taps into Internet servers through which almost all email traffic flows. Prior to the September 11 terrorist attacks, members of the U.S. Congress were critical of this invasion of privacy. After the attacks, there was little criticism.

How we evaluate these systems comes down to how we value control, security, freedom, and privacy. This issue comes up several times in the next chapter.

The hurrier we go,
the behinder we get.
— Anonymous

Progress vs. Obsolescence

The more we progress in our technology, the more technology becomes obsolete. The costs of obsolescence include environmental, social, and cultural.

Cellular telephones last between 16 and 18 months in the U.S., not because they break or become unusable, but because people switch to a new service that requires a different phone or because they simply want something new. Styles and features change as sizes diminish, making old phones conspicuously dated. It is a testament to the power of both technology and marketing that a device that did not exist a century ago now is obsolete in less than two years.

Progress can be seen in the sleek, ever more featured phones that ever more people carry with them. Obsolescence can be seen in the millions of phones disposed of. Containing toxic substances such as lead, cadmium, and mercury, obsolete cellular phones may, by 2005, constitute 65,000 tons of trash per year.

Obsolescence can cost jobs, too. Even before the 1960s, there were threats to end the bracero program, which brought thousands of Mexican workers into California. In response agricultural professors at the University of California at Davis developed a technology to compensate for the loss of labor in the harvesting of tomatoes. Their system combined mechanical engineering with genetic selection to create both a machine to harvest tomatoes and a tomato hard enough to come through without squishing. The combined impact of this technology and the political change in the bracero program could be seen in California between 1962 and 1970:

- The mechanical harvester and the hard tomato were born
- 1152 mechanical harvesters and 18,000 human sorters replaced 50,000 human pickers
- 600 growers replaced 4000
- Acres cultivated fell by 17% but production increased by 5%
- Mechanical harvesting rose from 0% to 99.9% of all production

The $65,000 mechanical harvesters were best suited for large farms, so those were the farms that benefited from the increased efficiency. When the bracero program did end in 1964, farms consolidated. The big got bigger and the small got out. Mechanical harvesting has helped California tomato production to pass the billion-dollar mark, but it also favored the large organizations. Technology progressed and the small farm and manual picker became largely obsolete.

In a story we explore in the next chapter, steel axes replaced the stone axes of an Australian aborigine tribe. This technological progress devastated the tribe's traditions and culture, which had long provided effective rules for living. Their world had changed and so did the rules for surviving. Their traditions and culture no longer fit the world that the tribe lived in. They had become obsolete.

The discovery of nuclear chain reactions
need not bring about the destruction of mankind
any more than the discovery of matches.
– Albert Einstein

Understanding the tradeoffs between technology's costs and benefits is becoming ever more important because, as technology's potential for benefit increases, so does its capacity for harm:

- Fire cooks and also burns.
- The crossbow defended and also conquered.
- Splitting the atom provided cleaner energy and also left near-immortal waste, as well as the power to annihilate the human race.
- Genetic engineering may eliminate most health problems and food shortages and it may also create global plagues against which we have no defense.
- Nanotechnology promises to provide fantastic material abundance and it could also wipe out all life on Earth.
- Robotics may support an evolutionary leap in human capability and it may also bring to extinction the race that spawned it.

We need tools to make such serious choices. The American Association for the Advancement of Science (AAAS) suggests the following questions in weighing the costs and benefits of technology:

1. What are alternative ways to accomplish the same ends? What advantages and disadvantages are there to the alternatives? What trade-offs would be necessary between positive and negative side effects of each?
2. Who are the main beneficiaries? Who will receive few or no benefits? Who will suffer as a result of the proposed new technology? How long will the benefits last? Will the technology have other applications? Whom will they benefit?
3. What will the proposed new technology cost to build and operate? How does that compare to the cost of alternatives? Will people other than the beneficiaries have to bear the costs? Who should underwrite the development costs of a proposed new technology? How will the costs change over time? What will the social costs be?
4. What risks are associated with the proposed new technology? What risks are associated with not using it? Who will be in greatest danger? What risk will the technology present to other species of life and to the environment? In the worst possible case, what trouble could it cause? Who would be held responsible? How could the trouble be undone or limited?
5. What people, materials, tools, knowledge, and know-how will be needed to build, install, and operate the proposed new technology? Are they available? If not, how will they be obtained, and from where? What energy sources will be needed for construction or manufacture, and also for operation? What resources will be needed to maintain, update, and repair the new technology?
6. What will be done to dispose safely of the new technology's waste materials? As it becomes obsolete or worn out, how will it be replaced? And finally, what will become of the material of which it was made and the people whose jobs depended on it?

If we apply this critical approach in great enough numbers then governments (in search of votes) and corporations (in search of dollars) will follow suit. But who are we to make such choices? Benefit and harm are in the eye of the beholder: Gutenberg's printing press spread knowledge and also the Protestant movement that undermined the Catholic Church. Since Gutenberg was a devout Catholic, from his standpoint this was a catastrophe. From Martin Luther's perspective this use of the printing press was a gift from God. This takes us from the technical issues of costs and benefits to the social and psychological issue of values.

Chapter 9

How Do We Evaluate Technology?

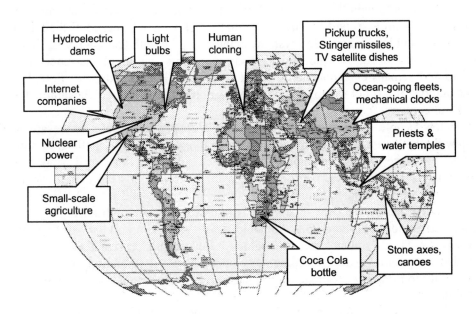

The easiest way to control a technology
is to kill it; then it can't possibly do any harm.
But if the public wants the benefits from the technology,
it must live with some risk and so must decide on
a tradeoff between the technology's benefits and its dangers.
Each society does this in a different way, depending on
its attitudes toward technology and its political culture.
— Robert Pool

One society adopts steel axes, but rejects canoes. Another uses Stinger missiles, AK-47 assault rifles, and pickup trucks, but bans televisions and satellite dishes. One country develops mechanical clocks and vast oceangoing fleets long before anyone else, but then destroys it all. When we evaluate technology we go beyond an objective weighing of costs and benefits to apply our own values. Technology's merit is in the eye of the beholder, so people with different values may agree on the costs and benefits of a technology, but come to entirely different conclusions on whether it is "good" or "bad."

There are at least two important reasons for us to understand how we evaluate technology. First, it improves our ability to communicate. After all, it is inevitable that we will interact with others whose values are different from our own, and we may have difficulty understanding their choices. We may even conclude that they are uninformed, stupid, or even malicious. The patterns in this chapter will make it easier to understand how rational, intelligent people can reach different conclusions when weighing the costs and benefits of technologies. Comprehending how we evaluate technology will also make it easier to explain our own conclusions.

The other reason to gain this understanding is that it helps us to reflect on our own values. Those who learn a second language say that it helps them understand their first language, and learning how other value systems come into play with technology is, like a second language, a mirror for us. Once we see them in contrast, we will better understand our own values and perhaps even question some. The ancient Greek philosopher Socrates said, "The unexamined life is not worth living." We might paraphrase him: "Unexamined values are not worth applying."

Just as few people think about technology on a conceptual level—as we are in this book—few think about their own values. With technology, we are not inclined to "reinvent the wheel," but rather use the technology that has already been developed or, perhaps, to improve it a little. With values, we follow the same approach. Instead of reinventing, we subscribe to the values of others: our parents, our social groups, and so on. We apply values that have proven themselves over time. This is practical because both technology and our values are part of countless activities: we could not ponder them hundreds or thousands of times each day. On the other hand, our values influence our present and future so profoundly that we cannot leave them completely unexamined. If technology is the lever that gives us ability to move great things, then values are the fulcrum that determines where the lever will move.

> We haven't formulated and agreed
> upon a way of making good decisions
> about the powerful technologies
> we're so good at creating.
> – Howard Rheingold

So how do we evaluate technology? The same way we evaluate anything in our lives. Psychologist Abraham Maslow suggested that our choices are driven by our needs, and some (such as the need for food and water) are primary. We are motivated to satisfy these physiological needs before all others, but once we do, we are free to concentrate on our next level of needs: safety and security. Then, once we satisfy those, we can focus on further needs—including belonging, love, and creative fulfillment. With each satisfaction of a more basic need, we are free to pursue satisfaction to "higher" needs. Maslow calls this our "hierarchy of needs" (also known as the humanistic theory of motivation):

1. **Physiological**: air, food & water, sleep, sex

2. **Safety & Security**: from current environment and future threat

3. **Social affiliation**: in family, occupation, religion, country

4. **Esteem & Self-esteem**: sense of usefulness and effect on environment

5. **Self-actualization**: fulfillment of ones potential (e.g. creative, physical)

No one suggests that we follow a strict step-by-step progression through the hierarchy. Rather, we strive to satisfy needs on multiple levels simultaneously. But if basic needs are threatened, then we tend to forget about higher needs until basic ones are once again satisfied. For instance, lacking physical security, it is difficult to work toward the intangibles of freedom. There are famous exceptions: Mahatma Gandhi fasted in protest to British injustice, sacrificing physiological needs in order to fulfill his potential as a leader toward a just society. Still, someone who declares, "Give me liberty or give me death," probably already has food, shelter, and physical safety...or believes that these will soon return.

The hierarchy of needs concept tells us that we evaluate technology as good or bad depending on whether it satisfies our most pressing needs, at whatever level those might be. If you were focused on survival, a big stick might be good. If focused on creative expression, acrylic paints or a word processor might be good. Everything else being equal, we choose the technology that helps us to satisfy our dominant needs. Usually, our needs are so obvious to us—so unconscious—that we are aware only of selecting the easiest option.

Ease is what the Coca Cola bottle offered the Bushmen of the Kalahari in the movie *The Gods Must be Crazy* (as we described in Chapter 6: "How does Technology Change Us?"). But when the bottle also brought discord, they evaluated it in terms of their mythology. The gods had always given them good things, never anything over which they had to compete. Receiving the bottle, therefore, must have been a mistake, and they evaluated it as bad.

A slightly different grouping of values is even better at explaining why groups and countries embrace or discard technology. As Maslow's hierarchy uses levels, Spiral Dynamics (developed by Clare W. Graves, Don Beck, and Jeff Cowan) uses "worldviews." In fact, Spiral Dynamics works so well that we adapt (with some liberty) six of its worldviews as answers to how we evaluate technology:

1. **Survival**: *Does it feed and protect us?* Robinson Crusoe gives us a fictional example of a shipwreck survivor on a desert island. When retrieving goods from his foundering ship, he selects pistol, saw, and hammer over gold because food and protection were most important.

2. **Ritual**: *Is it consistent with our myths and traditions?* The Yir Yoront aborigines in Australia evaluated steel axes as good and canoes as bad based on mythology. The steel axes introduced by missionaries were very similar to stone axes, which they believed the gods had given them. But gods had given them nothing like the canoes that their neighbors used.

3. **Power**: *Does it give us more control?* In Afghanistan, the Taliban evaluated Stinger missiles, AK-47 assault rifles, and pickup trucks as good because those maintained control, and television and satellite dishes as bad because those undermined it.

4. **Authority**: *Do our leaders or traditions say it is good?* China, guided by the authority of Confucian principles, developed mechanical clocks and vast oceangoing fleets of ships long before

Europeans did. But a new regime declared new priorities, scrapping much of the advanced technology and turning inward to build the Great Wall.

5. **Economic**: *Is it profitable?* In the capitalist free market system, electric lights, nuclear power, and Internet technologies promised profit and, so, were considered good. Labor-intensive, small-scale agriculture is less efficient than automated, large-scale agriculture, so the market puts the former out of business.

6. **Ecologic**: *Are environmental and long-term costs outweighed by the benefits?* The U.S. government evaluates hydroelectric dams based on how they affect water quality, fish, wildlife & botany, public recreation, historical & archeological sites, and aesthetics.

Of course there are countless ways to categorize human values and, therefore, countless ways we could explain how people evaluate technology. But, as in every chapter, our aim is to provide a context in which to understand the question, not declare which answer or set of answers is best. We are also not attempting to critique the Yir Yoront, Taliban, ancient China, the capitalist free market, or any other group or system.

These examples were selected because they illustrate diverse approaches to evaluating technology. By necessity, these brief descriptions of approach are generalizations. No group makes choices homogeneously; a host of details keep any evaluation from neatly fitting into any one category. But this book—and science and the pursuit of knowledge, in general—is about finding patterns. And these patterns are a useful reflection of what we see in the history of people evaluating technology.

Buddhist philosophy reminds us "the map is not reality." Good point. A map always sacrifices detail of the landscape it represents, and yet we still find maps useful. And so it is with our map of answers to how we evaluate technology.

Our map is descriptive, not prescriptive. It tells us how people *do* evaluate technology, not how they *should* evaluate it (attempts to impose value systems on other people have historically been bloody and largely unsuccessful). While this book does not preach one set of values to guide the future of technology, it does suggest that we educate ourselves to be able to answer these important questions:

- How do we, as a global population, make the complex and momentous decisions that are creating our future?

- How do we decide which technologies to develop and which to abandon?

- How do we decide which should receive more resources and which incorporate such danger that they should not be developed?

> I found out the carpenter's chest, which was indeed
> a very useful prize to me, and much more valuable than
> a ship-loading of gold would have been at that time.
> – Robinson Crusoe

Survival

In some situations, survival is our only concern. As babies we are unaware of much beyond food and warmth. And in many parts of the world, famine and war assure that survival remains the dominant priority. Other values seem luxuries for those struggling to get enough food to live another day or trying to avoid being caught in crossfire.

In Daniel Defoe's fictional story *Robinson Crusoe*, the namesake character is left on a remote island and must figure out how to stay alive. Crusoe evaluated technology based on whether it contributed or detracted from his ability to feed himself, stay warm, and stay safe. Crusoe's priorities were, "First, health and fresh water…Secondly, shelter from the heat of the sun. Thirdly security from ravenous creatures, whether men or beasts. Fourthly, a view to the sea, that if God sent any ship in sight I might not lose any advantage for my deliverance…" To him, more valuable than gold was technology that:

- Fed him (muskets, pistols, kegs of gunpowder, powder horns, bags of shot, knives, forks, bread, sugar, flour, and rum)

- Protected him from the elements (clothing, ropes, sail, canvas, hammock, bedding, razors, scissors)

- Protected him from "ravenous creatures, whether men or beasts," (saw, ax, hammer, bags of nails and spikes, screwjack, hatchet, and many of the hunting weapons in the first category)

Some of the tools, including a grindstone, could be used to make new tools, if he were not soon rescued. As it turns out, his priorities of food and shelter were appropriate because he was a castaway for a long time.

Of the costs and benefits we examined in the previous chapter, complexity vs. predictability would probably be the most important to those focused on survival. Complex technology may have a greater benefit over time, but its unpredictability would be too high a price. Long-term benefits do not matter if any short-term setback means you will not be around to enjoy them.

If the situation changes, either through experience (babies grow up) or reduction of threat (assured supplies of food), the individual or organization can then consider other needs and values.

> The stone axe in all its aspects,
> uses, and associations was
> integrated into the context of
> Yir Yiront technology and conduct
> because a myth, a set of ideas
> had put it there.
> — Lauriston Sharp

Ritual

The Yir Yoront, a nomadic tribe in Australia, had used stone axes for as long as any of them knew. Their mythology explained the entire world in terms of gods guiding their tribe since the beginning of time. Gods had given one ancestor the stone ax and all his descendants—his clan—shared the responsibility for honoring it. Another clan shared the responsibility for honoring dead ancestors. Each group, with their symbols or totems, knew its roles because of the mythological stories that were told again and again.

When Christian missionaries explored Australia mid-20th century, they brought new technology, including steel ax heads. From the missionaries' viewpoint, steel axes were good because they were more efficient, and they thought that "progress" in any form was good. By that time steel had been around more than two millennia, having been first smelted around 500 BCE in India.

Missionaries and tribe members alike agreed that steel axes were more effective than stone axes. But the Yir Yoront adopted steel axes as a good technology because they fit into the tribe's mythology. The only issue was whether they belonged to the totem of the stone ax, because they looked and worked like stone axes, or to the totem of dead ancestors, because the missionaries who brought them were as white as the sun-bleached bones of dead ancestors.

It may be tempting to project values of "efficiency and progress" on the process, assuming that these are universally compelling motivators. And from there we would conclude that mythology played only a superficial role in the Yir Yoront evaluation of stone axes. To determine whether mythology or efficiency was more influential, we consider the canoe.

The Yir Yoront did not use canoes. Instead, they held onto floating logs when paddling across rivers and lagoons—tricky business, as these were often infested with crocodile, shark, Portuguese man-of-war jellyfish, and stingray. As it turns out, the stingray was also important to the tribe since its barbed spine made a good tip for spears, catching in the wound and breaking off inside it. Those dried spines were the tribe's primary export and necessary for obtaining stone ax heads, the nearest source of which was hundreds of miles and many tribes away. So why did the Yir Yoront not use canoes?

A good guess would be that they had never seen a canoe and, so, simply had not come up with the concept. That would be wrong, as the Yir Yoront saw their neighbors to the north use canoes. Another guess might be that they lacked the raw materials to manufacture canoes and, while they could trade for small things like stone ax heads, trading for big technology was impractical. Also wrong. Trees suitable for canoes (and for making the handles and strapping of axes) grew in Yir Yoront territory. Their reason? The gods had never given them the canoe.

They assumed—well, they could imagine no other explanation—that the tribe to the north has a god that long ago gave them the canoe. Adopting canoe technology would not simply be a matter of learning how to build a canoe, but of creating an entire myth about how one of their ancestors had received it...and spreading that belief throughout the community.

So, efficiency—and utility, in general—played a secondary role in the Yir Yoront evaluation of technology. Primary was their mythology, which explained for them how the world worked and, so, how they should evaluate options and make decisions. For the Yir Yoront, gods and mystical spirit beings were quite real, so living according to the rules set down by those entities in myth made sense. That may not be scientific, but then neither is skipping the "unlucky 13th floor" in a skyscraper, numbering them 11, 12, 14, 15, and so on, as is common in the U.S. Just as some find safety in not working on the 13th floor of a building, others find it in preserving rituals and sacred places.

Impact of the Steel Ax

Oblivious to Yir Yoront traditions, the well-intentioned missionaries gave steel axes in exchange for work and also as gifts. The male elders, who had held power in part through control of the stone axes, were less eager to switch to steel axes—which were not under their control—than were younger members. The tables were turned as the male elders had to borrow from women and young people, undermining the stone ax as a symbol of masculinity and of respect for the traditional leaders. Also, the annual festivals where the Yir Yoront traded spears for stone axes with neighboring tribes lost their main purpose once they started getting steel axes. The ritual around this exchange lost its core meaning, attendance declined, and cultural interaction withered.

Liberation for the formerly dependent and underprivileged? By Western values, these changes were progressive, but they shattered culture, tradition, and even religious practices. Even if replacing a culture could be justified, the missionaries were unprepared to shore up the collapse and form a new society, and certain practices emerged that would not have been permitted before. For example, wives and daughters were prostituted to obtain use of steel axes. The social structures that would have prohibited such activities were unable to keep pace with the rate of change.

Were there any benefits? The missionaries hoped that the new technology would make the Yir Yoront more productive, and, so, raise quality of life. Instead, the efficiency allowed them to work less and further extend their sleeping hours, which the missionaries thought were already more than ample. The meaning of the ax was far more important in this society than was its function. Mishaps like this may have inspired the "non-interference directive" on the science fiction show Star Trek, which prohibited the technologically advanced "Federation" from disrupting more primitive cultures.

The Yir Yoront's evaluation process worked for them, or they would have perished as a tribe long before the missionaries arrived. Ritual is often based on practices proven effective and sustainable over long periods of time. Indonesian agriculture gives us an example.

The "Green Revolution" brought new agricultural techniques to Indonesia, much like the steel axe, these represented progress. The Indonesian government, who believed that progress was good, replaced the Balinese method of Hindu priests and water temple telling farmers when to irrigate their rice fields. The cascading of water from a high crater lake through a hierarchy of rice fields and dams (water temples, each with its own priest) appeared haphazard, inefficient, and unscientific. So the government imposed centralized and coordinated irrigation at more frequent intervals.

At first this had the desired effect of increasing crop yield, but then weaknesses became clear. The pattern of irrigation directed by the priests of the water temples had avoided the simultaneous ripening of all fields, which would have overtaxed water supplies. Progress provided rats and brown leaf-hoppers with contiguous fields of ripe crops across which they could multiply without limit.

The organochloride pesticide response to this was an expensive import and killed everything: fish, eels, and even farmers. The result was disastrous and the government had to allow the return of the ritual-based technique.

So what technique would have been an improvement? To figure that out, anthropologist Steve Lansing employed the latest technology: computer simulation of water flow, crop growth, and parasite spread. Working with the top priest, they repeatedly adjusted the timing of planting and irrigation to maximize yield. Their result was a pattern of irrigation very similar to what had been done, by ritual, for over a millennium.

So was the only problem in this case imperfect information? There is never perfect information—even 21st century science cannot explain and predict everything, and scientists will be the first to admit this—so when dealing with complex technologies, such as agriculture, it may sometimes be safer to stick with the tried-and-true approach…at least until we understand how and why it works.

While few societies today rely as much on mythology and supernatural forces in evaluating technology as the Yir Yoront or the Indonesian water temple priests, every society retains elements of it. Such beliefs are particularly common when the scientific mechanisms are hard to understand.

In Africa, the invisible and hard-to-predict behavior of AIDS prompts many to evaluate condoms, sexual practices, illicit IV drug use, and HIV drugs with values based on ritual. Overwhelmed by the complexity of the science surrounding AIDS, they rely on tradition, legend, and folk tales. But values evolved for one environment may not serve in another. For instance, one folk tale maintains that having sex with a virgin will remove the disease from ones body, and science can confidently say that this can only infect the virgin.

Tradition need not be based in mythology or ritual to exert strong influence. As we noted earlier, conscious reevaluation of *every* technology we encounter would be impractical—there is just not enough time in each day—so even those who practice science make many decisions based on tradition. For instance, modern business dress for men traditionally includes a necktie. Culture enforces this tradition subtly through peer pressure. As a clothing technology, the tie does little other than convey information ("I am conforming to our tradition" or "I am expressing my personality through the patterns on this tie").

Globalization brings ever more values systems into contact, as more and more technology is sold, traded, given, or imposed from one society to another. When such technology is not aligned with native values, clashes result. Avoiding these requires that we understand the ways we evaluate technology. Since every society contains a blend of the value categories we outline in this chapter, this knowledge is universally useful.

After the Mujahidin parties came to power in 1992,
the Afghan people thought that peace would prevail
in the country. However, the leaders began to fight
over power in Kabul. Some local leaders,
particularly in Kandahar, formed armed gangs that
fought each other. There was widespread corruption and
theft, and there were roadblocks everywhere.
Women were being attacked, raped, and killed.
—Taliban Spokesman

Power

Shiny black tape was wrapped around a pole in the middle of the street. Like the medieval practice of putting decapitated heads on pikes outside the castle walls, it was a warning. The entrails of cassette music tapes warned not about trespassing, however, but about technology.

In Afghanistan, Taliban doctrine prohibited satellite dishes, television, videotapes, music tapes, and even traditional Afghan instruments and singing. Kites, which centuries earlier had led to a revolution in electrical technology, were also prohibited by the doctrine. All that flew in the breeze was the confiscated magnetic tape.

Pickup trucks, AK-47 assault rifles, and Stinger missiles (shoulder-mounted devices capable of shooting down helicopters and airplanes) were accepted technologies. Evaluating technologies of power and control as "good" made sense, since the Taliban came to power in the chaos of feuding warlords that followed a decade of Soviet occupation. In that environment, survival required power, either your own or that of a group that would protect you. Power was the only currency both in the anarchy the Taliban replaced as well as in the strict order they imposed.

Hungering for stability, some Afghans accepted the strict and repressive new regime. Others joined the Taliban for protection. Many more were simply conquered by force, burying their televisions and other forbidden technology in their backyards (to be unearthed after the fall of the Taliban in late 2001).

Trading freedom for stability reflected in how the Taliban evaluated technology. Television satellite dishes could bring in opposing views and were, therefore, subversive of the regime and considered bad technology.

Totalitarian regimes in other countries and eras have also evaluated technology based on "might makes right." For instance, North Korea. While the U.S. continues its long debate over registration of guns—and how that might threaten the right to bear arms guaranteed by the U.S. Constitution—North Korea registers radios. Any citizen who owns a radio must register it with the local police, and any foreign-made radio must also have its tuner soldered into place, stuck at the official frequency. To ensure that no one modifies a radio to pickup anything other than official government programs, police make surprise inspections of the radios.

The reason for North Korean government's paranoia? Radio Free Asia broadcasts news on North Korea's shortages of food and electricity, as well as its political isolation. North Koreans know that they are hungry but few are aware that others are better off, as government propaganda has been consistent and complete since before many citizens were born. Radio Free Asia also broadcasts instructions on how to escape the country, including how to dress and act, and whom to contact. Successful defectors have been surprised to find that South Korea is wealthier than North and that the U.S. is not subservient to North Korea, but donates rice to it for reasons other than tribute.

Evaluating technology rather differently, human rights activists in South Korea smuggle small, disposable radios across North Korea's borders. The smaller the radios are, the easier they are to hide from police. The more disposable they are, the more likely that there will be nothing for the police to find when they do search.

North Korea experimented with mobile phones for almost two years, deciding May, 2004, to ban them. Tens of thousands subscribed even though the cost to register a phone (about $750 U.S.) equaled 28 years of the average worker's wages. Mobile phones, too, allowed foreign culture into the country, undermining government control.

The life experiences of those living in violent, oppressive environments confirm and re-confirm that you have to fight for whatever you get, that allegiance is based on power and force, and that anyone who does not believe this will suffer the consequences—and may not have the chance to change their mind. This "survival of the fittest" mentality is not re-stricted only to violent, ruthless environments. It is a worldview we can find, at least in pockets, anywhere on earth.

The North Korean government's evaluation of technology is not that different from the Taliban's. Technology that promotes freedom undermines control, and technology that promotes control undermines freedom. Given the examples of this cost-benefit tradeoff in the previous chapter, many would concur, but while analysis of tradeoffs can be fairly ob-jective, evaluating the tradeoffs rarely is.

> In the Islamic value system, there is not the same emphasis on individ-ual freedom. The individual is seen as absorbed within, and subject to, society…Energy is more likely to be put into group efforts to improve society than into individual effort to pursue an individual course.
> – Peter Marsden

Authority

Although many Taliban decisions appear based on the value of power, they aspired—at least in word—to the higher authority of the Koran. They claimed to be creating a pure Is-lamic state. While *power* focuses on the individual (How does this serve me?), *authority* subordinates the individual to a greater good (How may I serve it?).

Taliban adherence to the Koran provides a different explanation for their ban on televi-sion. Islam prohibits visual representations of human and animal form. During Islamic rule of Spain, art and architecture contained brilliant geometric patterns, completely devoid of human or animal images. What could be worse than a painting with human or animal forms? An electronic box that presents a variety of them 24 hours a day.

Other religions also influence how their followers evaluate technology. For example, many Catholics oppose human cloning because the Vatican condemns it. They view the

Pope as interpreting the authority of the Bible and conveying the authority of God. In contrast to the other worldviews we have discussed, they do not evaluate human cloning based on immediate survival (I need it now or I may perish), ritualistic tradition (we have always manipulated our world), or power (I can dominate others with this technology). They evaluate it based on whether it is God's will for us (concluding that it is not right to meddle in this area).

Authority and power can blend since people in positions of authority may evaluate technology based on how it protects and enhances their power. They may even believe that a higher authority would be best served if they kept power. For them, the claim of authority may be a self-serving deceit, and they may even deceive themselves. Their followers may believe strongly that those in authority are acting purely in service of the ideology, whether religious, political, or something else.

The power of authority had far-reaching consequences in 15th century China, where it led to England colonizing Hong Kong instead of what might have happened: China colonizing Ireland or other parts of Europe. This happened because China evaluated ocean-going ships as bad and a protective wall as good. Based on the values of avoiding contact with foreigners and celebrating agriculture over commerce, it came at a time when China had by far the largest and most sophisticated fleet of ships in the world.

The roots of this decision trace back to the 5th and 6th centuries BCE, a violent and brutal period in China. Victors of battles sometimes slaughtered tens or hundreds of thousands of civilians. In this environment, a philosopher named Confucius praised the importance of authority and the stability of the state. Evaluating based on "what is righteous" rather than "what brings power" created a more structured society. For those who suffered through the many wars of ancient China, sacrificing individualism for order was probably quite attractive.

Confucianism values relationships over the individual. Or, more specifically, it defines the individual in terms of his or her relation to parents, children, siblings, friends, ruler, and subordinates. Carrying out ones social roles properly is of utmost importance. This philosophy dominated China for nearly two millennia, but in the 15th century a combination of power and economic values guided the emperor. This resulted in China's "Treasure Fleets," so named because they exchanged Chinese treasure (e.g. porcelain and silk) for foreign treasure (e.g. spices and gold).

Early in the 15th century, well before Europe discovered the New World, China explored as far as India and Africa with fleets of up to 300 ships—some of them five times the size of the ships that Christopher Columbus sailed to discover the Americas. Not only larger, but also more technologically advanced, Chinese ships were divided into separate watertight sections. Perhaps patterned after the structure of bamboo, they were constructed so that no single breach of the hull would sink the ship. This design would not appear on European ships for another three centuries.

In addition to building the largest, most technologically advanced fleet of the era, China also introduced a string of inventions: paper, printing, gunpowder, the mechanical clock, and the magnetic compass. Europe, which was splintered into many small countries, provinces, and city-states, would have had little defense if confronted with China's vast fleets and advanced technology. But Europe had nothing that China wanted.

The mind of the superior man
dwells on righteousness;
the mind of the little man
dwells on profit.
— Confucius

And then Confucians retook power, evaluating technology according to their traditional values, which elevated the farmer above the merchant. Farmers must have been relieved because they had been taxed mercilessly by the eunuch regime to build and stock the fleets of ships. The new rulers dismantled the fleets, shut down shipbuilding factories, and eventually made it against the law to sail on the ocean. They also destroyed other technology of their predecessors, including mechanical clocks.

Confucian values shunned foreigners, an almost instinctive reaction in 15th century China, when stories of Mongol hordes invading overland were still fresh. Kubilai Khan, grandson of Ghengiz Khan, had conquered China and Mongol control continued from 1279 until 1368. With that history, a defensive wall on the frontier must have seemed a much better technology than fleets of ships. The wall had been started 1600 years earlier under the Qin and then the Han Dynasties, but was vastly expanded in the 15th century under the Ming Dynasty at the expense of all exploration.

Isolation and Shock

What made China's decision to abandon naval technology so serious? No strong neighbor continued developing shipping when China dropped it, quickly exposing the mistake. Instead, it took four centuries before China could compare its choice against the alternative.

China was ready to start an industrial revolution centuries before Europe got around to it, but political unity made possible their decision to stop development of technology. Europe's political fragmentation made it impossible to focus the massive resources China did, but also made it impossible to abandon development of new technology. If one country—say, Holland—developed oceangoing fleets, then other countries—say, Spain and England—would note Holland's success and copy it. Countries that ignored successful examples would become marginalized and, sometimes, overrun.

Naval technology shrank the world, eventually making China and Europe neighbors. The consequences of China's 15th century choice became clear in the middle of the 19th century, when European ships (armed with cannons using Chinese-invented gunpowder) demanded access to her ports for trading. The technological disparity was so stark that China had little defense and, so, was exploited economically.

In the second half of the 20th century China again evaluated technology according to authority. Communist authority stopped technological progress during the Cultural Revolution of the 1960s, but the disadvantages were stark and quickly obvious. The "freeze" was much shorter this second time around because China was no longer isolated.

In the 15th century it must have been difficult to predict that wall technology would be so much less valuable than ship technology. In the 21st century many independent organizations and countries are exploring many different technologies, and since transportation and communication technology has connected them all, any missed opportunities are quickly exposed.

Authority guides many of the people alive today, whether manifested in religion or politics. It defines what is right and wrong, good and bad. But many sources of authority date back centuries or millennia, and could not have anticipated specific technology. It is to the followers of authority to analyze how ships, clocks, radio, television, cloning, and countless more inventions fit within the teachings. This can be complicated and even contradictory (e.g. cloning could save precious lives even as it dares to create life). But there is another way to analyze technology, and because it is more quantitative than qualitative, its results are less subjective.

> Anything that won't sell,
> I don't want to invent.
> — Thomas Edison

Economic

Thomas Edison claimed that an invention that did not turn a profit was worthless. A century before Edison invented the light bulb, Adam Smith speculated that capitalism and the free market worked as if an "invisible hand" were making decisions, setting just the right prices to make demand equal supply. By encouraging competition the invisible hand would keep prices as low as possible. Taking this approach to evaluating technology, we need ask only if it would make money. More broadly, we might ask if it contributes to greater economic efficiency, progress, or growth.

Asking those questions is the starting point. Answering them involves weighing costs and benefits for the known alternatives (which include doing nothing at all) by *assigning economic value* to the effects of the alternative, *assessing risk*, and *discounting the future* (balancing future benefit against present benefit). The alternative with the greatest economic value wins.

Edison's light bulb, for example, produced brighter light than candles and reduced the risk of fire, but it consumed electricity. We can place an economic value on electricity because it has a market price. Because candles do not *always* cause fires, assigning a value is harder. Insurance companies place economic value on such occurrences using historical statistics to determine the likelihood of various types of fires and the cost of repair.

But what is the value of brighter light? Is it the profit from manufacturing and commerce that could not have been conducted with just candlelight? Is it the improved productivity of workers who have not damaged their eyesight by straining in dim light? Or the saved medical expenses of treating them (assuming some treatment existed and that they would be treated)? Or is the value of brighter light simply whatever the market would pay for it?

Even if that is the case, our cost-benefit analysis is not complete. We have to consider acquisition costs for light bulbs and candles, and we have to make some assumptions about the relative value of something now versus something in the future. To switch candles to light bulbs we would have to purchase bulbs, switches, and wiring, connect to an electric utility, and establish a contract for purchasing electricity at a predictable rate.

Because these costs are immediate but the benefits accrue over time, determining whether the investment is worthwhile depends on where else you could place your money. If interest rates were very high, you might earn more by keeping money in the bank than investing in light bulbs and their infrastructure. A $1000 investment might save $10 per year if spent on light bulbs and infrastructure, but earn $100 per year if placed in a bank.

Once we assess all these factors, we could conclude whether a light bulb creates more economic value than its alternatives. If so, it is a good technology. But, technically, our conclusion is not guaranteed to be right. The future is unpredictable, so either wax or electricity might become much less (or more) expensive, or interest rates might (always do) change. And that would render our cost-benefit analysis invalid.

But it is an imperfect world and our cost-benefit analysis may be good enough. Plus, the market makes continual corrections, motivating people in new directions, so this economic approach to evaluating technology is adaptable and resilient. For instance, if electricity shot up in price, the market would provide a profit motive for development of high-efficiency light bulbs or the reintroduction of candles.

All of this balancing of options and present against the future is complicated when the technology in question does not even exist yet. When evaluating an investment in a prospective technology, venture capital firms assess three dimensions:

1. The *product* itself (Are there barriers to entry of competitors, such as patents or specialized knowledge?)

2. The *people* proposing to deliver it (Are they experienced? Do they have a good track record?)

3. The *market* for the product (Does it exist? Has it been receptive to new technology?)

In some cases, the market values growth over profit. Market participants—entrepreneurs and businesses—recognize that securing a lion's share of the market can lead to large long term profit, and so sacrifice a smaller immediate profit. In the bull market of the 1990s, investment capital flowed to companies with plans to dominate Internet purchases of pet food, groceries, airline tickets, gift certificates, on-line greeting cards, toys, or furniture.

Focused on growth and promises that profits would eventually come in lavish amounts to those companies that locked-in their market segment, investors plowed their money in. But lock-in and profit are not as easy to achieve as raw growth.

The lock-in that Microsoft had achieved with their operating systems was because customers moving to a new operating system had to undertake the expensive and risky process of replacing applications and migrating data. Internet buyers tended to have little loyalty to commerce web sites because they could switch from Pets.com to Petopia.com every time they bought a bag of dog food, patronizing whichever competitor was offering the biggest loss-leader discount.

With time, few "dot com" companies found either lock-in or profits, so many closed down. Conventional, bricks-and-mortar pet supply stores acquired both Pets.com and Petopia.com. In the end, economic evaluation says that profitable technology is good and unprofitable technology is bad.

> You could say that capitalism is just
> a couple-hundred-year-old mechanism
> for speeding up science, but
> capitalism and the free market
> are not very good at saying
> "pause," let alone "stop"
> – Bill Joy

Ecologic

A young Chinese woman wearing a red smock and black pants, but no eye protection, swings a hammer at the base of a television picture tube. For years, the glass body braced against 14 pounds of air pressure on every square inch, protecting the vacuum inside. The hammer changes that in an explosive instant. If all goes well, only the tail (yoke) of the tube breaks free and air rushes into the void. If all does not go well, the glass can shatter and implode. Either way, once the prized coils of wire are stripped from the yoke to sell as bulk copper, the lead and barium laden glass is dumped.

The tubes—technically, cathode ray tubes or CRTs—are "good" technology in Guiyu, China, where they are evaluated by strict economics: cost of acquisition, transportation, and labor subtracted from the market price of copper. But these CRTs have become "bad" technology in the U.S., where they were used and discarded by people upgrading to larger televisions or to flat panel computer screens. To recycle a CRT costs between $15 and $40, and some states, including California, prohibit dumping them in landfills because of their toxic contents, which can leach into soil and ground water.

Ecological evaluation of technology considers environmental factors that economic evaluations traditionally ignore. These real but unaccounted costs include impacts to health and the environment long into the future. Traditional economic evaluation "externalizes" costs, such as lead contamination in groundwater. Ecologic evaluation considers factors like the lead-poisoned fish and the people that have few, if any, other sources of food.

In the case of the CRT, the lead once served an important function: protecting the viewer from high-energy electron radiation firing at the inside of the screen to illuminate it. Ecological evaluation would weigh this health benefit against the cost of safe disposal and any alternatives to the CRT. This was not done when and where CRTs were made or sold, so when they came to the end of their useful life, the alternatives were proper recycling, which nobody had committed to—or wanted to—pay for, or "offshore recycling," which meant dumping in countries where labor is cheap and environmental protection is minimal.

Environmental laws level the playing field for competing corporations, so those that protect the environment are not under priced—and eventually eliminated—by those that do not. While individuals may see ecologic evaluation as simply the right thing to do, corporations are called on to maximize stockholder value, and that can be served by externalizing as many costs as possible. If corporations' directors do not drink the local water or eat the local fish, dumping may be attractive because it is cheap.

One justification for government is to protect that which is held in common. If the factory next door to you pollutes its own air, it also pollutes yours. If a hydroelectric plant upstream from you dams its river, it also dams yours. In such a position, you would likely want government to enforce ecologic evaluation of technology. We find an illustrative example in the Pacific Northwest.

The Snake River flows west across Idaho to Oregon, heading north into Washington before resuming its westward travel to the Pacific Ocean. Along the way, it passes through Hells Canyon, a river gorge averaging more than a mile deep. In 1955, the U.S. government issued the Idaho Power Company a license to generate electricity by damming the river. Idaho Power built Brownlee Dam in 1958, Oxbow in 1961, and Hells Canyon in 1967. Together known as the Hells Canyon Complex, these three hydroelectric dams can produce more than 1.17 million kilowatts of power, about 1.6% of the developed hydroelectric generating capacity in the U.S. (In 2001 hydroelectric provided 6% of the nation's electricity).

The license to operate these three dams expires in 2005, 50 years after its 1955 issuance. To continue operating, Idaho Power needs a new license from the Federal Energy Regulatory Commission (FERC), which evaluates technology on factors beyond economics. So, in order to gain FERC's approval, Idaho Power is conducting a decade-long, multimillion-dollar study, proposing to spend approximately $178,000,000 on:

- Water quality
- Fisheries
- Wildlife and botanical protection and reintroduction
- Public recreation access, facilities, and condition
- Historical/archeological preservation
- Aesthetics

In exploring "water quality," Idaho Power is looking at more than just pollution. If the temperature of water coming off a dam is too high or low, fish can be hurt or killed. The amount of oxygen dissolved in water affects aquatic life in the top two meters, and that, in turn, affects the rest of the ecosystem. Depending on how they deflect or spill water, the dams can increase or decrease oxygen content. Keeping water still in reservoirs reduces it and spilling water through the air, much as a waterfall would, increases it.

Dams block anadromous fish (those that migrate from saltwater to freshwater for spawning, such as salmon and steelhead) from returning upstream to their breeding grounds. One solution is to provide a way around the dams (e.g. stepped pools), but at the Hells Canyon Complex this approach encountered problems, so hatcheries were built downstream of the dams instead. Since anadromous fish sense their way back to their birthplace, those born in hatcheries return there, rather than upstream of the dams where their ancestors may have bred. Conservation groups maintain that hatcheries address only one aspect of the disruption to the fish and the greater environment.

Constructing and operating a technology such as the Hells Canyon Complex affects wildlife habitats. Narrow channels of water are converted to reservoirs, and cold, fast water slows down and warms in the sun, affecting fish, aquatic plants, animals that feed on them, and the habitats that those animals reach. Even the hatcheries cause a chain reaction leading to less fertilization of soils, which affects trees (as we mentioned in the Overview section at the beginning of this book). So part of the relicensing process includes proposals by Idaho Power to improve habitats adjacent to the complex. One example: reintroduction of mountain quail. Another is the protection of sensitive plants and the control of invasive weeds.

And there are human considerations. Long before the dams were built, and before European settlement of the Pacific Northwest, tribes of American Indians roamed and lived in the area. Preserving archeological sites and enhancing them with interpretive centers is part of the overall evaluation. Impact on current tribes (Burns Paiute, Nez Perce, Shoshone-Paiute, Shoshone-Bannock, and the confederated tribes of the Warm Springs Reservation and Umatilla Reservation) is also a factor, so Idaho Power is proposing projects and programs to benefit them.

On top of that, Hells Canyon Complex shares the Snake River with recreational boaters, fishers, and hikers, and campers. Since it will be evaluated on its impact on these recreational users, Idaho Power proposes to improve litter and sanitation programs, improve road maintenance, enhance the four Hells Canyon parks, and improve or expand boat docks, ramps, and launches, among other projects. Aesthetic improvements to the Hells Canyon Complex also influence the likelihood of re-licensing, so Idaho Power proposes to blend facilities, structures, and signage into the landscape.

So, while Idaho Power evaluates the Hells Canyon Complex on economics, the U.S. Government evaluates it ecologically on these six areas: water quality, fish, wildlife & botany, public recreation, historical & archeological sites, and aesthetics. The government protects the Snake River and surrounding Hells Canyon Recreation Area on behalf of the nation's citizens, allowing a private enterprise to profit from it, provided that enterprise performs on certain non-economic measures.

Other technologies exploit common resources with less ecologic evaluation. Combustion of coal to generate electricity creates airborne particles that fall onto bodies of water. One pollutant, mercury, converts to an even more toxic compound, methyl mercury, in the environment. Methyl mercury bio-accumulates up the food chain, becoming more concentrated at each step. Large fish such as tuna can concentrate enough that it may be safe for humans to eat no more than one 6.5 ounce can per week.

The U.S. government studies and evaluates the health risks of such toxins as mercury, passing laws limiting output of these pollutants. Instead of ecologic evaluation, these external costs may be ignored until regulation sets a certain "free" level, above which manufacturers or users of the technology must pay in the forms of fines or lawsuits.

Pharmaceutical technologies are evaluated the other way around, assessing risks and costs before release of product. But it was not always this way. Prior to government regulation, untested drugs were sold from the backs of wagons. Society recognized the devastating and irreversible harm that can take place when not all repercussions of a drug are well understood, and insisted on government oversight. That recognition is spreading to technology beyond drugs. Weighing the possible costs, both direct and ecologic, against the alternatives prior to releasing the technology is called the *Precautionary Principle*. It places the burden of proof on those creating the new technology to document its impact in comparison to alternatives, which include sticking with existing technology. The Precautionary Principle is very important for those who are not involved in selecting a technology because they often bear the brunt of the costs.

Benefits are often anticipated and directly affect those selecting the technology. Costs are often unanticipated and indirectly affect many uninvolved in selecting the technology. Competition and self-interest leads to this. Someone inventing, developing, and selling a technology will succeed, at least in the short term, by focusing on those who might purchase it. For them, benefits must outweigh costs. Whether the technology succeeds in the marketplace is much less dependent on those only peripherally affected by it, so for them benefits need not exceed costs. It is also simply easier to focus on and study a small group of direct beneficiaries than the diverse and possibly dispersed population of those indirectly affected.

One way we already protect our diverse and disperse environment is with deposits on recyclable beverage containers. These bottles and cans are less likely to be dumped because they can be redeemed for cash. Although the pennies may be insufficient to change behavior of many consumers, some people retrieve discarded containers because they rely on the deposits. The container deposit is represented as the value of the material for recycling, but it could also be considered the cost to our environment of dumping the material.

Economist Robert Costanza of the University of Maryland combines the beverage container deposit concept with another already in general use: construction performance bonds. In the U.S., construction companies commonly post bonds guaranteeing completion of a major project. Incomplete or late projects cause all or part of the bond to be forfeited. Costanza advocates the Precautionary Polluter Pays Principle ("4P"), which would require developers of new technology to post an interest-bearing bond to cover a worst-case scenario. An independent scientific study would determine just how costly the proposed technology could be. Over time, if the technology does not cause environmental damage or can be shown to be safer than initially predicted, portions or all of the bond with interest are returned to the developer.

4P would make the developer responsible for covering ecologic costs of a technology, rather than forcing those affected by it—but not profiting from it—to prove it is harming them. We noted in the last chapter that some asbestos victims died before courts found in their favor because the legal process can take years or decades.

Although 4P is attractive in its simplicity, it leaves difficult questions. What bond would be appropriate for technology that could have irreversible impact? When would a bond be liquidated for nuclear technology, whose wastes remain dangerous for many thousands of years? What would a bond pay for if a genetic technology changed humans? Who would collect the bond if a technology were to make the human species extinct, as a global exchange of thermonuclear weapons threatened to do? In the extreme, the Precautionary Principle dictates that the risks outweigh any bond that could be posted and the technology should not be pursued.

Ecologic evaluation considers those who are not directly involved with the technology. It takes a long-term approach, asking whether a technology is sustainable and how it will affect future generations. Long-term thinking is becoming more important with increasingly powerful technology, and considering the full context may help us avoid extinction. Narrow economic evaluation can be myopic: Lenin once pointed out that a capitalist would sell rope to his own hangman.

If Bacteria Used the Precautionary Principle

Give a little DNA. Get a little DNA. Bacteria are pretty easygoing about evolution—none of that formal male-female sexual reproduction. As a result of their sharing snippets of genetic instructions, plus a bit of random mutation, bacteria have explored and adapted to almost every environment on this planet, from deep inside rocks to steam fissures on the bottom of the ocean to the human intestine. But what if bacteria were not so cavalier about trying out new forms? What if they had the intelligence and consciousness to employ the Precautionary Principle to their evolution?

If there were just one evolutionary creation that bacteria could have prevented it would surely be cyanobacteria (more commonly known as blue-green algae, even though it's not an algae). More than 2.2 billion years ago, a mutation allowed cyanobacteria to get the hydrogen atoms they consumed from water, while other bacteria continued to get their hydrogen from sugars, airborne hydrogen, or hydrogen sulfide. With water immensely abundant, cyanobacteria multiplied almost without bounds, and produced a lot of waste. Because they consume H_2O and use hydrogen, their waste is oxygen.

From less than one billionth of one percent of the atmosphere, oxygen skyrocketed to its current level of 20%. All life on earth had evolved to that point with virtually no free oxygen, so this change was a global disaster. Volcanoes and asteroid impacts were mild in comparison. Even humans, whose lineage evolved in an oxygen-rich environment, suffer from oxidation, and may take anti-oxidant supplements (e.g. vitamins C, E, and beta carotene) to counter oxygen's tendency to react with and change a variety of compounds.

Bacteria were—and are—not in a position to employ the Precautionary Principle, but humans are. What might we learn from the cyanobacteria episode? One lesson is that something that can both replicate itself and consume an untapped food or energy source will change the world. We may eventually be able to create our own "cyano-technologies" that brilliantly exploit some previously untapped resource, much as the runaway nanotechnology does in Michael Crichton's fictional novel *Prey*.

With technology changing at an accelerating pace and with surprising behaviors emerging from new complexity, we may well create a variety of technologies with the potential to change the world. If we would like to survive—and not have to hide under a rock, side by side with the bacteria still waiting for oxygen to go away—then this is a fine time to become familiar with the Precautionary Principle...and start evaluating our technology ecologically.

Ours is a progressively technical
civilization...committed to the quest for
continually improved means
to carelessly examined ends.
– Robert K. Merton

One way or another, consciously or not, we all evaluate technology. Usually we do it unconsciously. The values on which we weigh technology are so much a part of us that we are unaware of them. If we were aware of what was going on, we might hear the following:

- A malnourished peasant: "Of course we evaluate based on whether it helps us get food."
- An aborigine in Australia: "Of course we evaluate based on what our spirit ancestors provided."
- A warlord in Afghanistan: "Of course we evaluate based on whether it maintains or expands our power."
- An imam in a mosque: "Of course we evaluate based on the teaching of the Koran."
- A partner in a venture capital firm: "Of course we evaluate based on profit."
- An environmentalist: "Of course we evaluate based on sustainability and compatibility with the surroundings."

And they all might add, "Anyone doing differently is a fool."

How do we reconcile these diverse and seemingly incompatible approaches to evaluating technology? Concerned that the average person is unable to understand the growing complexity of the technology on which he or she relies—and, therefore, unable to predict or choose its future impact—author James Burke offers four alternatives:

1. Return to an intermediate technology
2. Assess scientific and technological research according to worth for society
3. Direct research and development towards more durability and less planned obsolescence
4. Allow technology to continue to evolve as it's always done

In the first case, we get a simpler environment, safe from the vastly powerful technologies recently developed. But what technologies are we willing to relinquish? Do we sacrifice life-saving technology? Few would be willing to surrender even the conveniences—which so quickly seem necessities—such as email and cellular phones. Our civilization is so interconnected, like a spider's web, that removing *any* strand affects the others. And removing *many* strands could be disastrous.

Burke's second alternative presents the immense problem of figuring out who will decide the social value of each technology. As we saw in this chapter, the various value systems make very different evaluations. Socialist societies have attempted to impose a centrally directed order with, often, unfortunate and self-destructive results. Those who direct scientific and technological research will have power, and power can corrupt.

But even if we could figure out some democratic system to distribute that power to everyone (and considering what is popular on television, creating a technologically-informed and participatory population would not be easy), there is a larger problem. Simply evaluating an area of research does not tell us the impacts of the eventual technologies. Intel's research into a better way to build a handheld calculator was not intended to create the microprocessor found in microwave ovens, antilock brakes, automated teller machines, and home computers. What was the worth to society, in 1969, of reducing the number of integrated circuits in a calculator? No one could have guessed.

Burke's third alternative is similar to our category of "ecologic." This compromise would reduce the energy focused on relentless progress and novelty, redirecting it to the less fortunate people in the world. The rich would stop wanting so much more new technology and the poor could get enough for at least their basic needs, and perhaps even catch up with the rich. As Burke himself points out, this is a utopian vision, but perhaps utopia will be feasible as it has not been in the past. Advances in agriculture and manufacturing technology make it much easier to create enough of life's necessities. Advances in communication and information technology make it much easier to share such a vision on a global basis.

Whether such a vision would work or not, we do know that the present system has created at least a trickle of advances down to the poor. And this is the last of Burke's four alternatives: allowing technology's evolution to continue as it always has. Although a fairer redistribution might be a noble goal, many have much more faith in "the rising tide lifting all boats." They believe that the best way to bring the benefits of technology to the poor is to let the rich forge ahead, following their natural, even selfish, desires.

This notion does have a track record. For example, lifespan in poor countries today matches what it was in rich countries a few centuries back. Luxuries common in the 21st century—for example antibiotics or cellular phones—could not have been had by emperors or kings even one century back. And consider the centralized Socialist/Communist economies of the 20th century. Despite their original vision, they did little to prove that meddling in the affairs of progress could be successful.

But what if just letting technology progress "naturally" leads to dire consequences? That is precisely what worries Bill Joy, someone who deeply understands technology. A child prodigy, Joy went on to study electrical engineering. He rewrote the Unix computer operating system while at UC Berkeley and was a founder of computer manufacturer Sun Microsystems, becoming its chief scientist.

Joy's concerns stem from three emerging areas: genetic technology, nanotechnology, and robotic technology (GNR). Not only could each of these technologies be designed to self-replicate, but economic and efficiency factors could motivate us to make them self-replicating. For instance, a company that would manufacture nanobots may be able to do so with less expense if it enlists the nanobots it has already made in order to make more. In this way, making just a few would result in a never-ending supply.

And, unlike nuclear technology, which relies on hard-to-get materials like plutonium, these three technologies depend mostly on information. Since information is so hard to control (witness the illegal copying of music on the Internet), it could eventually become quite feasible for a bright, but discontented person to create and unleash a technological plague.

Combining elements of authority, economics, and ecologic, Joy offers a five-pronged proposal to prevent this from happening:

1. Compel scientists, technologists, and corporate leaders to pledge they will not work on weapons of mass destruction (a variant of the Hippocratic Oath)
2. Create an international organization similar to the U.S. Office of Technology Assessment (which was closed in 1995) to evaluate new technology
3. Make corporations financially liable for the consequences of the technology they develop
4. As was proposed for nuclear technology in 1946 (but not adopted), internationalize control of technologies evaluated as too dangerous to be developed commercially
5. Renounce development of technology so dangerous that it could escape our control and threaten our entire species

Joy's detractors claim that his worry is unjustified, that technology has long increased in power, often appears too powerful, and yet always works out to improve our lot. They would say that Burke's laissez faire fourth alternative is not only the best choice, but the only workable one simply because pursuit of knowledge cannot be stopped. When it is stifled in one place (e.g. China destroys its fleets of ships in the 15th century), it simply goes somewhere else (e.g. Europe, whose fleets came to dominate the world). So they claim there is nothing to worry about and, even if there were, there's nothing we could do about it anyway.

> The views of scientists should not
> have special weight in deciding
> questions that involve ethics or risks:
> indeed, such judgments are best left to
> broader and more dispassionate groups.
> — Martin Rees (scientist)

But stifling technology is not the only alternative to the public giving a blank check to the experts directing technology. Those experts wager not just their own lives, but ours, too, so even if we do not understand the technical details of their projects, we may want a say in what values they employ and what risks they term "acceptable." Given the radically diverse ways in which people evaluate, we may wish to question the experts' entire process of evaluation.

Recognizing the stakes, the U.S. Congress amended a 2003 bill funding nanotechnology with a provision requiring "regular and ongoing public discussions, through mechanisms such as citizens' panels, consensus conferences, and educational events…" Average citizens may not be experts on nanotechnology, but they are experts on their own values. The process of evaluating technology can benefit from technical understanding, but it is absolutely dependent on values.

What would it look like to ask citizens to evaluate technology? Denmark has asked their citizens to evaluate such technological issues as electronic surveillance, noise and technology, genetically modified food, infertility, and the future of the private automobile. "Consensus conferences" start out like jury duty, with the Danish Board of Technology sending out invitations to randomly selected citizens. Of those interested, the Board selected a panel of 14 based on diversity of age, gender, education, profession, and geography. The panel receives a brief technical education over two weekends and then spends three days at the conference. For one and a half days experts answer questions from the panel and from a general audience. Then the panel develops a report, striving to achieve consensus before the report is publicly presented. After the presentation, the experts correct any factual misunderstandings and the report is submitted to the Danish Parliament.

Key to the success of citizens' panels is unbiased technical information. Or, failing that, multi-biased technical information. Just as the U.S. Congress is subject to the sway of special-interest groups providing expert testimony skewed to serve their own political objectives, so, too, would be citizens' panels. Two factors that might reduce the influence of skewed information: (1) testimony is given in a public forum and (2) citizens are not seeking campaign contributions from the employers of the experts.

Switzerland calls citizens' panels "PubliForums" and France calls them "Citizens' Conferences." Every country involving citizens in the this way faces the challenge of giving their citizens a tool to understand and evaluate technology. For making conscious choices based on both a broad, contextual understanding and on examined values, ICE-9 would be a good tool.

The Tao of Technology

We join spokes together in a wheel,
but it is the center hole
that makes the wagon move.
We shape clay into a pot,
but it is the emptiness inside
that holds whatever we want.
We hammer wood for a house,
but it is the inner space
that makes it livable.
We work with being,
but non-being
is what we use.
— Lao-Tzu

Was Lao-Tzu describing the context of technology? Writing the Tao Te Ching, basis for the Taoist philosophy, 27 centuries ago, he described the importance of what surrounds technology, pointing out that it takes meaning from relationships just as we do. In the 21st century, the quickening pace of change forces us from the comfort of focusing on narrow aspects of our world to recognizing the panorama of its interconnectedness.

Viewed another way, technology is one thread in the tapestry of our world. It touches and affects every other thread, and is touched and affected by them. So, if we aim to understand our world, technology is a good thread to follow. And now, at the dawn of the 21st century, is a particularly important time to understand our world. Technology has given us great leverage over our world and its future, but the lever swings both ways.

In the previous chapter, we mentioned that Bill Joy was concerned about genetic technology, nanotechnology, and robotic technology for two reasons. First, its creation depends more on information (which happens easily, as the music recording industry discovered with Napster and its file-sharing successors) than on rare materials (e.g. plutonium). And, second it can self-replicate. Earlier, in the chapter on how technology works, we described the possibility of nanotechnology, in the form of nanobots, replicating out of control like a cancer. In fact, Joy sees market forces toward "better and improved" products making such technology inevitable unless we approach them with a view of the larger context.

Viewed myopically, each technical advance from current technology to self-replicating nanobot could make good, economic (or military) sense. Who would not want something a little better than the tools we have now? Power and profit motivate their development. Each step would make sense to the corporations or governments funding it, but the total product could make any power or profit gained utterly meaningless. Nanobots that could scavenge atoms from the air around them (performing the beneficial service of removing carbon pollution and, so, reducing global warming) could scavenge from anywhere.

Whether through unfortunate error or malicious intent, nanobots could build their progeny from the atoms in soil, plants, buildings, and even our bodies. They would convert matter in any form they found it, and as with flu virus, killing some of the nanobots would not stop their advance. As we saw with distributed organization in the chapter on how technology works, technology can be as resilient to component failure as an ant colony is to individual ant failure. Even one remaining nanobot would make another, and then each would make another. As we saw with exponential growth in the chapter on how technology changes, this quickly results in huge numbers.

We have created immensely powerful tools and we may find compelling reasons to develop in the direction of technology that we would be unable to control. What alternative do we have to locking our doors and hoping that "the experts" will make the right decisions for us? Within our grasp is a rich, contextual understanding of technology and a considered, critical approach to evaluating it. Using that, we will be able to choose the sort of future we want and then to influence the technological decisions toward that direction. So this book is not about just an interesting new way to view technology, but an important way to cope with the rapid change it causes so that we can guide that change.

> For education the lesson is clear:
> its prime objective must be to increase
> the individual's 'cope-ability'—the speed
> and economy with which he can adapt to
> continual change.
> – Alvin Toffler

In *Future Shock*, Alvin Toffler writes about the stress of cultural and technological change on humans. He recognizes that this rate of change is new to human civilization and that it requires response from education. One response is this book. Another is the school curriculum we mentioned in the chapter on how we change technology, which presents the same ICE-9 approach (one lesson per chapter) on a different level. Whether we use ICE-9 or another strategy for understanding and evaluating, we need an approach that lifts us above the blur of detail that so many equate with technology. We need the sort of enduring and lucid patterns that we have found in this book:

What is technology? It is applied science and it also precedes science. It is the tool that extends our abilities, the system of tools and techniques, and the information that underlies all of this (much as DNA coding underlies life).

Why do we use it? We want food, shelter, communication, transportation, commerce, art, religion, health, entertainment, organization, exploration, and some even want conflict. These applications existed in antiquity, now motivate the latest inventions, and will effectively categorize future technology.

Where does it come from? One technology leads to another, by both extending our physical ability to create and demonstrating that something more is possible. In populations made dense by proximity or telecommunication, we share our developments, triggering further innovation. Specialization allows us to pursue refinements full time. Both plan and accident create technology. Protection of intellectual property can both stimulate and inhibit innovation.

How does it work? Every technology needs energy, often converting it from one form to another. Organization can be centralized or distributed, with each having advantages. Control systems use feedback and correction to accomplish a goal. The information controls technology, and in the form of algorithms it may determine behavior independent of how the technology is implemented. Mechanisms and processes may be repeated many times and/or incorporated within other layers of mechanisms and processes. New behavior, unpredictable from any of the component parts, emerges from certain organizations of technology.

How does it change? Technology becomes inconspicuous by becoming common, and it disappears by becoming incorporated into other technologies. Necessity is not only the mother of invention, but also its daughter when brilliant inventions cast about for worthy applications. When a technology offers advantages, confers prestige on its adopters, is compatible with existing systems, and all this is clearly visible before risking its adoption, then many will take that risk. The process of autocatalysis describes how technology acts on itself to change itself, which often leads to accelerating change. That accelerating change is also characteristic of the process of evolution, which is not restricted to biology passing information through genes, but applies to technology passing information through memes.

How does it change us? Improved nutrition, sanitation, and medicine have nearly quadrupled human lifespan, in spite of our dense and connected populations offering better environments for disease. Spear, plow, factory, ship, and computer have changed how we work, rendering some occupations obsolete while creating new. Books, newspapers, television, and the Internet change what we consider to be true, just as drugs and virtual reality change what we consider to be real. But, even beyond our perceptions, technology changes our very species by changing the environment in which both technology and we are evolving, perhaps on a collision course.

How do we change it? Fortunately, we are not fated to passively observe technology changing us. We consume some technologies and let others fail in the marketplace. We invent new, improving on the past. We govern with both prohibitions and subsidies as politicians, judges, and voters. We manage the countless decisions that guide development and diffusion of technologies. We invest the resources necessary to advance technologies. We also question and reevaluate them. And we teach about technology in general and these roles in particular.

What are its costs and benefits? Independent of the specific technology, we encounter some universal tradeoffs. The more useful something is, the more dependent on it we become. The more capabilities and features we demand in new technologies, the more complex and less predictable they become. Costs and benefits can be sudden (catastrophic) or slow (chronic), and we regularly trade one for the other and back again. Technology can provide us more control and security in our society, but usually at the cost of freedom...and it can provide more freedom at the cost of control. The wonder of ever-improving technologies leaves a trail of obsolete artifacts and skills, and the faster they improve, the more they render worthless.

How do we evaluate it? Beyond the objective weighing of costs and benefits lie our highly subjective values. If focused on survival, we might evaluate a technology on whether it feeds and protects us. If ritual is central, then we ask if it is consistent with our myths and traditions. Where power rules, does it enhance and protect our clout and control? Those who subordinate themselves to a higher authority ask if their leaders or traditions deem it good. Economically motivated ask, "Is it profitable?" Those focused on ecology ask, "Are its benefits greater than its environmental and long term costs?"

These are not the best answers. They are a launching point for each of us to look for our own answers. They set a context for the ICE-9 questions and they set a standard that future answers will need to meet or beat. They make some answers appear glib and superficial—such as technology being computers or coming from factories.

And, once we have found effective coping responses, we need to spread these skills so that others will also thrive in this rapidly changing technological world. Education has never been so urgent, so share this book. Give it to teachers and parents you know, and tell them about the KnowledgeContext curriculum. Find new answers as well as new questions, and share those (with this author, too). As we noted at the beginning of this book, creating an intentional future is a collective process.

> Human history becomes more and more
> a race between education and catastrophe
> – H.G. Wells

For perspective, we close with two modern sages: Arthur C. Clarke and Carl Sagan. Clarke's *2001: A Space Odyssey* (also a movie by Stanley Kubrick) opens three million years ago on a harsh and dangerous land where wild cats top the food chain. Our forebears, more ape than man, discover an alien monolith, smooth and shiny, which imparts to them the idea of using tools. With bones as levers, they strike down prey, fight off predators, and battle each other. Over time, the bones are copied and improved into a world of technology, which mankind uses to go beyond our world. On the moon, astronauts in the year 2001 discover a second monolith, buried beneath rock and dust. Uncovered, the monolith signals its alien creators that mankind has evolved to a technological threshold. We have evolved.

Let us imagine a different ending to Clarke's story. In our version, the alien civilization shares the concern that Carl Sagan voiced in the quote opening this book: that our "world-altering contrivances," unlike the bones from which they sprang, pose a threat to our civilization's very existence. Having proven that we can use tools, but not that we can do so consciously and wisely, we might need help. So, instead of sending an announcement to extraterrestrials, the second monolith would show us the step that follows the ability to use technology. Inscribed with something akin to the ICE-9 questions, the monolith would impart the idea of understanding and evaluating.

Clarke's book is science fiction, not prophecy. The year 2001 has passed and we have found no monoliths, either signaling faraway aliens or presenting an alien rendition of ICE-9. But we have found something smaller and more convenient than a monolith: this book.

Going Beyond this Book

Introduction

Carl Sagan's quote about the time of perils
> Sagan, C. (1994). Pale Blue Dot. Random House, New York. (p. 371)

Concern that a nuclear detonation would set our planet's atmosphere ablaze
> Goodchild, P. (1985). Shatterer of Worlds. Fromm International Publishing Corporation, New York. (p. 54)
>
> Stoff, M.. et al (1991) The Manhattan Project: A Documentary Introduction to the Atomic Age. Philadelphia: Temple University Press.

Some of the ways technology and science could extinguish us
> Rees, M. (2003). Our Final Hour. Basic Books, New York.

Institute for Biological Energy Alternatives synthesizes completely new virus
> Pearson, H. (14 November 2003) Virus built from scratch in two weeks: New method accelerates prospect of designer microbes. Nature Journal.
> http://www.nature.com/nsu/031110/031110-17.html

Smallpox resistant to all known treatment. Viruses and bacteria that degrade the human immune system or modify behavior:
> Davis, C., Orent, W. (18 March 2004) Smallpox: Scourge of the Past, Weapon of the Future? British Broadcasting Corporation:
> http://www.bbc.co.uk/science/genes/gene_safari/wild_west/smallpox.shtml

Dangers of genetically-engineered bioweapons
> Schram, M. (2003) Avoiding Armageddon. Basic Books, New York)

Fictional account of future nanotechnology and artificial intelligence run amok:
> Crichton, M. (2002). Prey. HarperCollins Publishers, New York.

Factual account of how nanotechnology has developed
> Regis, E. (1995). Nano the emerging science of nanotechnology: remaking the world—molecule by molecule. Little, Brown and Company, Boston.

How nanotechnology might change our world
> Drexler, K. E., Peterson, C., Pergamit, G. (1991). Unbounding the Future: The Nanotechnology Revolution. William Morrow, New York.

Overview

Martin Rees' quote "...there could be a crucial hurdle..."
> Rees, M. (2003). Our Final Hour. Basic Books, New York. (p.164)

How North Korea evaluates communication technology
> Brooke, J. (3 March 2003) Threats and Responses: Airwaves; Infiltrators of North Korea: Tiny Radios.
> http://query.nytimes.com/gst/abstract.html?res=F70F11F634580C708CDDAA0894DB404482
> Reuters (22 August 2003) Police Block Attempt to Send Radios Into North Korea. New York Times Newspaper. http://www.nytimes.com/2003/08/22/international/asia/22WIRE-KORE.html.

Interfax (6 April 2004) North Korea bans mobile phones.
http://www.interfax.com/com?item=Chin&pg=0&id=5728986&req=

How hydroelectric dams affect salmon and, thereby, the forests
Tolan, S. (25 August 2000) Point of No Return: Salmon Saga. Living on Earth, WBUR Radio.
http://www.loe.org/archives/000825.htm (with a link to listen to show over the Internet)

Chapter 1: What Is Technology?

Hawaiian bobtail squid use bioluminescent bacteria to hide from predators
Pearson, H. (13 December 2001). Blame It on the Bugs. Nature Journal.
http://www.nature.com/nsu/011220/011220-2.html

Hawaiian bobtail squid produces reflectin, inspiring new technology
Crookes, W. (9 January 2004). Reflectins: The Unusual Proteins of Squid Reflective. Science
Magazine (www.sciencemag.org/)

Tactics of vibrio cholerae in the human stomach
Ewald, P. (1996) Evolution of Infectious Disease. Oxford University Press, New York.

Essays on the nature and history of technology as the history of humans
Lienhard, J. (2000) The Engines of Our Ingenuity: An Engineer Looks at Technology and Culture.
Oxford University Press, New York. http://www.uh.edu/engines/

Do birds use technology?
Pain, S. (8 August 2002) Crow reveals talent for technology. NewScientist.com news service.
http://www.newscientist.com/news/news.jsp?id=ns99992651

Homer's Iliad gives techniques for launching and landing ships
Hobart, M. (1998). Information Ages: Literacy, Numeracy, and the Computer Revolution. Johns
Hopkins, University Press. (p. 19)

Touchstones and the origin of money
Burke, J. (1978). Connections. Little, Brown and Company, Boston. (pp. 16 – 17)

Chapter 2: Why Do We Use Technology?

19th century telegraph resembles 20th century Internet
Standage, T. (1998). The Victorian Internet. Walker and Company, New York.

Relationship of primate neocortex with size of social group
Dunbar, R. (1996). Grooming, Gossip, and the Evolution of Language. Harvard University Press,
Cambridge, Massachusetts.

Israeli tapping of Iraqi telephone lines
Hamza, K. (2000) Saddam's Bombmaker. Scribner, New York. (pp. 183 – 184)

Secret communication through the ages (our quote comes from page 1)
Singh, S. (2001). The Code Book. Delacorte Press, New York

Cigarette technology
Krock, L. (October 2001) Anatomy of a Cigarette. Nova PBS
http://www.pbs.org/wgbh/nova/cigarette/anatomy.html

Technology helps the blind to see

> Hall, A. (11 December 2000) Sonar Sight for the Blind? BusinessWeek Magazine.
> http://www.businessweek.com/bwdaily/dnflash/dec2000/nf20001211_136.htm

Using technology for social dating

> Horn, D. (29 January 2004) Fugitive done in by savvy date and Google. The Cincinnati Enquirer.
> http://www.usatoday.com/tech/news/2004-01-29-google-bust_x.htm

Chapter 3: Where Does Technology Come From?

Both dense populations and specialization play into a theory of how win-win relationships contribute to increasing complexity (and technology)

> Wright, R. (2000). Non-Zero: The Logic of Human Destiny. Pantheon Books, New York

Influence of war and the industry that creates technology for war (including quote by Solly Zuckerman)

> Zuckerman, S. (1982). Nuclear Illusions and Reality. HarperCollins Publishers, New York

Inventors patent ideas to preempt their rivals

> Pimental, B. (9 June 2003). Offensive Blocking Patents. San Francisco Chronicle.

James Burke quote "Either history is a series of individual and unrepeated..."

> Burke, J. (1978). Connections. Little, Brown and Company, Boston. (P. 287)

How many technologies were invented long before is commonly believed

> James, P. and Thorpe, N. (1994) Ancient Inventions. Ballantine, New York.

Chapter 4: How Does Technology Work?

Perpetual motion machines and why they cannot exist

> Lemonick, M. Will Someone Build a Perpetual Motion Machine? Time Magazine.
> http://www.time.com/time/reports/v21/science/motion.html

Sources of energy include wood and Ethiopia uses few alternatives

> Thurow, R. (26 November 2003). Unable to Tap Power of the Nile, Ethiopia Relies on Fuel Carriers. Wall Street Journal.

Reindeer pull sledges in northern Europe about 7,000 years ago

> Schobert, H. H. (2001) Energy and Society. Taylor & Francis Group, New York. (p.15)

Where does U.S. energy come from?

> U.S. Department of Energy (www.ne.doe.gov)

Foundation for 16 acres of wheat + millions of years = 1 gallon of gasoline

> Mason, B. (29 October 2003). Calculations illustrate fossil-fuel crisis. Nature Journal
> (http://www.nature.com/nsu/031027/031027-3.html)

In the sun's fusion reaction, matter is converted to energy according to Einstein's famous e=mc2 equation at the rate of 4,600,000 tons per second. Of the energy released from the sun, only one part in two billion reaches the earth:

> Schobert, H. H. (2001) Energy and Society. Taylor & Francis Group, New York.

Centralized vs. decentralized power generation over the years

> 177 power stations in the United Kingdom power stations as of 2002:
> http://www.dti.gov.uk/energy/inform/energy_stats/electricity/dukes5_10.xls (+ some smaller than

1MW, and renewables some of which are larger than 1 MW, but are not broken out). Total power in MW: 73,455 + 5,407 (renewables) = 78,862.

Life in 11th century England from the Domesday Book: http://www.domesdaybook.co.uk/

Eniac cost $486,804.22 in 1945 dollars equal to about $5,119,557 in 2004 dollars. What is a dollar worth? These websites help to calculate the value of dollar at different times

http://woodrow.mpls.frb.fed.us/research/data/us/calc/ and
http://www.orst.edu/dept/pol_sci/fac/sahr/sahr.htm.

Protein folding calculated by > 200,000 CPUs

http://folding.stanford.edu/

Emergent behavior in nature and technology through engaging stores

Kelly, K. (1994) Out of Control: The Rise of Neo-Biological Civilization. Addison-Wesley, New York.

Read more about ant colony behavior

Holldobler, B. and Wilson, E. O. (1994) Journey to the Ants: A Story of Scientific Exploration. Harvard University Press, Cambridge, Massachusetts.

Chapter 5: How Does Technology Change?

Puts technological change into context of the Universe evolving ever faster

Kurzweil, R. (1999) The Age of Spiritual Machines: When Computers Exceed Human Intelligence. Penguin Group, New York.

Air speed doubled every nine years (and land speed almost as quickly)

Lienhard, J. (2000) The Engines of Our Ingenuity: An Engineer Looks at Technology and Culture. Oxford University Press, New York. Pp. 142 – 143.

Why efficiency will lead to technology replacing biology as dominant life form:

Moravec, H. (1999) Robot: Mere Machine to Transcendent Mind. Oxford University Press, New York.

Musings and reflections of a high-tech executive flying around the world

Cochrane, P. (1999) Tips for Time Travelers: Visionary Insights into New Technology, Life, and the Future on the Edge of Technology. McGraw-Hill, New York.

Gutenberg's press printing Bibles, prayer books and papal indulgences

Burke, J. (1985). The Day the Universe Changed. Little, Brown & Company, Boston (p. 116)

Setting up to print a book costs 3x having a scribe copy it

Burke, J. (1985). The Day the Universe Changed. Little, Brown & Company, Boston (p. 113)

More about the Eniac computer

Story of: http://ftp.arl.mil/~mike/comphist/eniac-story.html
Tube failure (1/19000) every 7 minutes: http://www.tcpd.org/mccain/handouts/FutureEdge.pdf

Symbiogenesis. The quote "Animal evolution resembles the evolution of machines" is on page 172.

Margulis, L. and Sagan, D. (2002). Acquiring Genomes: A Theory of the Origin of Species. Basic Books, New York.

How does technology move from invention to widespread adoption? What are the human factors in technology' success and failure?

Rogers, E. M. (1995). Diffusion of Innovations. The Free Press, New York

Chapter 6: How Does Technology Change Us?

Lifespan, nutrition, suicides, and homicides
> Simon, J. (1995). The State of Humanity. Blackwell, Cambridge. This book contains:
> - Grantham, G. Agricultural Productivity Before the Green Revolution [lifespan & nutrition statistics]
> - Haines, M. Disease and Health through the Ages [Michael Haines quote from page 56]
> - Chesnais, J. Trends in Historical Murder and Suicide [Suicides & homicides p.96]

Mortality and life expectancy
> U.S. Census www.census.gov/prod/2002pubs/01statab/vitstat.pdf
> Center for Disease Control (CDC): www.cdc.gov/nchs/data/hus/tables/2002/02hus028.pdf
> Unicef: www.childinfo.org/cmr/revis/db2.htm
> United Kingdom statistics on suicides and homicides
> http://www.statistics.gov.uk/cci/nugget.asp?id=185

Information about Karachi, Pakistan including dumping of untreated sewage
> United Nations www.irinnews.org/report.asp?ReportID=19178
> CIA Factbook www.cia.gov/cia/publications/factbook/geos/pk.html

50-75 trillion cells in human body and ranges of cell lifespan
> Carnegie Library of Pittsburgh. (1994). The Handy Science Answer Book. Visible Ink Press, Detroit, Michigan. (p. 268)

Causes of leukemia
> http://www.oncologychannel.com/leukemias/causes.shtml.

Telomeres and life extension
> http://www.sciam.com/article.cfm?articleID=0009BC68-36E7-1CE5-93F6809EC5880000&pageNumber=1&catID=9

Faeroes measles epidemic and much more on the development of civilization
> Diamond, J. (1999). Guns, Germs, and Steel. W.W. Norton & Company, New York (p. 203)

Flush toilet of 1884
> Burke, J. (1996). The Pinball Effect. Little, Brown & Company, Boston (p. 139)

Stimulating pleasure center in brain of a rat to control its actions
> Hoag, H. (19 June 2003). Neuroengineering: Remote Control. Nature Journal (www.nature.com)

Use of rhymes to assist memorization prior to writing
> Burke, J. (1985). The Day the Universe Changed. Little, Brown & Company, Boston (p. 99)

Genetic evidence for divergence of wolves and dogs
> PBS Nova Online http://www.pbs.org/wgbh/nova/wolves/03false.html

Robert Wright quote "This is history's drift..."
> Wright, R. (11 September 2003). Two Years Later, a Thousand Years Ago. New York Times. www.nonzero.org/nyttwoyears.htm

Marshall McLuhan quote "First we build tools, then they build us"
> Smart, J. (2003) Accelerating Times: Signs of the Singularity. http://singularitywatch.com/newsletter_issues/atimes1-8_2003p2.htm

Chapter 7: How Do We Change Technology?

A variety of surprising ways in which people have—often unintentionally—changed the course of technology
> Burke, J. (1978). Connections. Little, Brown and Company, Boston.

Emperor Vespasian dismissing a mechanical invention: "You must let me feed my poor commons."
> Rolfe , J. C. (1914). Suetonius / Vespasian 18. The MacMillan Co., New York.
>> Also online: http://www.fordham.edu/halsall/ancient/suetonius-vespasian.html

Why did the light water nuclear reactor win in the U.S.? The typewriter's story. Juanita Ellis questions nuclear power plant
> Pool, R. (1997) Beyond Engineering: How Society Shapes Technology. Oxford University Press, New York.

How government continues to change nuclear power technology
> Hebert, H. J. (10 June 2003). Senate OKs Billions for Nuclear Power. Associated Press. http://www.electricityforum.com/news/jun03/senatenuclear.html

The many roles people played in influencing personal computer technology
> Freiberger, P. and Swaine, M. (2000) Fire in the Valley: The Making of the Personal Computer. McGraw-Hill, New York.

How invention and diffusion of technology is more soap opera than laboratory
> Flatow, I. (1992) They All Laughed...From Light Bulbs to Lasers: The Fascinating Stories behind the Great Inventions that Have Changed Our Lives. HarperCollins Publishers, New York

How does public policy change technology?
> Wenk, E. (1995). Making Waves: Engineering, Politics, and the Social Management of Technology. University of Illinois Press, Urbana, Illinois.

Chapter 8: What Are Technology's Costs And Benefits?

Horse pollution in New York (Converted 1 acre x 175' to 200x200x190' based on 43,460 ft^2/acre.)
> Dwyer, J. (1989). Strange Stories & Amazing Facts of America's Past. Reader's Digest Association, Pleasantville, New York.

Compare accident statistics for horse-drawn vehicle and cars
> O'Brien, E. and Benedict, L. (2001) 19th and 20th Century: Horse, Wagon and Automobile Accidents. Chicago Public Library's Municipal Reference Collection. http://www.chipublib.org/004chicago/disasters/19th20th_accidents.html

Aftermath of a nuclear accident analyzed by National Institutes of Health
> *Long Term Follow-Up of the* Residents *of the Three Mile Island Accident Area: 1979-1998*
> *http://ehp.niehs.nih.gov/docs/2003/5662/abstract.html?section=environmental*

Dumping of electronic waste from U.S. to Asia
> Basel Action Network and Silicon Valley Toxics Coalition. (2002). Exporting Harm: The High-Tech Trashing of Asia. www.ban.org/E-waste/technotrashfinalcomp.pdf

Use of radio to aid slaughter in 1994 Rwanda genocide
> Burstein, D. and Kline, D. (1995). Road Warriors: Dreams and Nightmares Along the Information Highway. Penguin Group, New York.

Death rates per mile traveled are 100 times higher for driving than flying
> UK (per billion kilometers) 0.02 air, 2.8 car http://plus.maths.org/issue15/editorial/
> U.S. (per 100 million miles) 0.003 air, 0.83 car http://www.nsc.org/lrs/statfaq.htm

Jonathon Philips, manager of DARPA's Human ID at a Distance program, quote, "We develop the technology. The policy and how you implement them is not my province."
    (September 2001) Big Brother Logs On. MIT Technology Review Magazine
    http://www.technologyreview.com/articles/01/09/mag_toc.asp

By 2005, INFORM Inc., an environmental research organization predicts that about 130 million cell phones weighing about 65,000 tons will be "retired" annually in the United States
    (8 May 2002) Inform: Strategies for a Better Environment.
    http://www.informinc.org/pr_wireless.php

International cellular phone usage by country
    CIA Factbook. http://www.cia.gov/cia/publications/factbook/fields/2151.html

Complexity vs.Predictability: new systems are so complex that accidents may be expected or "normal"
    Perrow, C. (1999). Normal Accidents: Living with High-Risk Technologies. Princeton University Press, Princeton, New Jersey

What patterns underlie major technological disasters?
    Evan, W., Manion, M. (2002). Minding the Machines: Preventing Technological Disasters. PrenticeHall PTR, Upper Saddle River, New Jersey.

What are the causes and consequences of technological breakdowns?
    Chiles, J. (2001). Inviting Disaster: Lessons from the Edge of Technology. HarperCollins Publishers, New York.

High-risk technology and human error intersect in a "calculus of catastrophe"
    Dumas, L. (1999) Lethal Arrogance: Human Fallibility and Dangerous Technologies. St Martin's Press, New York.

How can we make the software that runs our hardware more reliable?
    Peterson, I. (1995) Fatal Defect: Chasing Killer Computer Bugs. Random House, New York

Why software cannot be made reliable enough for critical systems
    Weiner, L. R. (1993) Digital Woes: Why We Should Not Depend on Sofware. Addison Wesley, New York.

How do we assess risk? Are we worrying about the wrong technologies?
    Lewis, H. W. (1990) Technological Risk. W.W. Norton & Company, New York.

Making technology safer can result—and has—in it becoming riskier
    Tenner, E. (1996) Why Things Bite Back: Technology and the Revenge of Unintended Consequences. Alfred A. Knopf, New York

48 technologies or technology-related issues rated according to likelihood of adverse effect and the severity of that effect
    Ropeik, D., Gray, G. (2002) Risk: A Practical Guide for Deciding What's Really Safe and What's Really Dangerous in the World Around You. Houghton Mifflin Company, New York.

Chapter 9: How Do We Evaluate Technology?

Maslow's hierarchy explained
    (2004) Abraham H. Maslow's Hierarchy of Needs. Accel-Team.com.
    http://www.accel-team.com/maslow_/maslow_nds_02.html

Yir Yiront aboriginal tribe's adoption of steel axes, but not canoes (Quote from p. 81)
> Spicer, E. (1952). Human Problems in Technological Change. John Wiley & Sons, New York. This book contains the essay Steel Axes for Stone Age Australians by Lauriston Sharp.

Peter Marsden quote on Islam and the Taliban
> Marsden, P. (2000). The Taliban. Oxford University Press

CRTs (ready to smash) and other e-waste exported from U.S. to Asia
> Puckett, J., et al (2002) Exporting Harm: The High-Tech Trashing of Asia. Basel Action Network and Silicon Valley Toxics Coalition. http://www.svtc.org/cleancc/pubs/technotrash.htm

Precautionary Principle
> Montague, P. (16 April 2004). Answering the Critics of Precaution. Environmental Research Foundation. www.rachel.org/library/getfile.cfm?ID=378.

Mercury in the environment and its health effects
> Clarkson, T. (February 2002). The Three Faces of Mercury. Environmental Health Perspectives. http://ehp.niehs.nih.gov/members/2002/suppl-1/11-23clarkson/EHP110s1p11PDF.pdf

Cyanobacteria wreaks havoc on environment (pp. 81 – 83)
> Margulis, L. and Sagan, D. (1995). What is Life? Simon & Schuster, New York.

Dangers of nanotechnology, biotechnology, and robotics
> Joy, B. (April 2000). Why the Future Doesn't Need Us. Wired Magazine http://www.wired.com/wired/archive/8.04/joy.html
> Joy, B. (24 May 2000). Welcome to the Age of Killer Technologies. The Age http://theage.com.au/

Alternatives to evaluating technology
> Burke, J. (1978). Connections. Little, Brown and Company, Boston. (p. 293)

Martin Rees' quote "The views of scientists should not have special weight ..."
> Rees, M. (2003). Our Final Hour. Basic Books, New York. (p.78)

Evaluating technology according to justice, participatory freedom, and economic development?
> Barbour, I. (1993). Ethics in an Age of Technology. HarperCollins Publishers, New York.

How does America evaluate technology and should it?
> Wenk, E.. (1986). Tradeoffs: Imperatives of Choice in a High-Tech World. Johns Hopkins University Press, Baltimore, Maryland.

Citizens' Panels and Consensus Conferences
> Danish Consensus Conferences website
> http://www.tekno.dk/subpage.php3?article=468&toppic=kategori12&language=uk
>
> Co-Intelligence Institute: http://www.co-intelligence.org/P-citizenCC.html
>
> International Institute for Environment and Development http://www.iied.org/)
> Report on Swiss and French versions: http://www.iied.org/docs/pla/pla_fs_14.pdf

Index

Acknowledgements

This book is more coherent and accurate than it otherwise would have been because of many generous reviewers. Mike O'Sullivan was relentless in making sure that the stories I told in one chapter were consistent with those I told in another. For extensive proofreading and critique of concepts, I thank Martina Flach-Aznar, Ignacio Aznar, Charles Hanes, Toni Rango, Trevor Oelschig, Dom Lindars, Chris Vaughan, Addie Holsing, Erin Murphy, Robyn Juster, Gene Ressler, Jim Puckett, Karl Pister, John Smart, and Robert Wright. Errors that persist are the fault of the author.

Lee LaFollette suggested Spiral Dynamics as framework for how we evaluate anything, including technology. Rob Mitrey helped determine the costs of producing just one microprocessor. Don Jacobs used the curriculum that parallels this book, testing it in middle school classrooms and giving feedback that has influenced the contents of this book. Charles Hanes, Mark Cathcart, and Trevor Oelschig helped to fund the printing book.

About KnowledgeContext

KnowledgeContext, a 501(c)(3) nonprofit corporation, helps young people think critically about technology. KnowledgeContext believes that all children should be offered the tools for thriving in a world where technology is changing how we study, what careers are available, the issues on which we vote, what we buy, and how we take care of ourselves.

Learn more at http://knowledgecontext.org

About the author

Miguel F. Aznar is Executive Director of KnowledgeContext. He is passionate about teaching how to understand and evaluate technology. Mr. Aznar discovered this calling after studying electrical engineering and computer science at the University of California at Berkeley and working as a software engineer (NASA, Amdahl) and management consultant (Ernst & Young, AT&T). He speaks publicly on this issue of technological literacy at conferences and on the media.

Contact him at aznar@knowledgecontext.org